# Environments

## FOSS
**Next Generation**

**Full Option Science System**
Developed at the Lawrence Hall of Science, University of California, Berkeley
Published and Distributed by Delta Education

**FOSS Lawrence Hall of Science Team**
Larry Malone and Linda De Lucchi, FOSS Project Codirectors and Lead Developers
Kathy Long, FOSS Assessment Director; David Lippman, Program Manager; Carol Sevilla, Publications Design Coordinator; Susan Stanley, Illustrator; John Quick, Photographer
**FOSS Curriculum Developers:** Brian Campbell, Teri Lawson, Alan Gould, Susan Kaschner Jagoda, Ann Moriarty, Jessica Penchos, Kimi Hosoume, Virginia Reid, Joanna Snyder, Erica Beck Spencer, Joanna Totino, Diana Velez, Natalie Yakushiji
Susan Ketchner, Technology Project Manager
**FOSS Technology Team:** Dan Bluestein, Christopher Cianciarulo, Matthew Jacoby, Kate Jordan, Frank Kusiak, Nicole Medina, Jonathan Segal, Dave Stapley, Shan Tsai

**Delta Education Team**
Bonnie A. Piotrowski, Editorial Director, Elementary Science
**Project Team:** Mathew Bacon, Jennifer Apt, Sandra Burke, Tom Guetling, Joann Hoy, Angela Miccinello, Megan Posco

**Thank you to all FOSS Grades K-5 Trial Teachers**
Heather Ballard, Wilson Elementary, Coppell, TX; Mirith Ballestas De Barroso, Treasure Forest Elementary, Houston, TX; Terra L. Barton, Harry McKillop Elementary, Melissa, TX; Rhonda Bernard, Frances E. Norton Elementary, Allen, TX; Theresa Bissonnette, East Millbrook Magnet Middle School, Raleigh, NC; Peter Blackstone, Hall Elementary School, Portland, ME; Tiffani Brisco, Seven Hills Elementary, Newark, TX; Darrow Brown, Lake Myra Elementary School, Wendell, NC; Heather Callaghan, Olive Chapel Elementary, Apex, NC; Katie Cannon, Las Colinas Elementary, Irving, TX; Elaine M. Cansler, Brassfi eld Road Elementary School, Raleigh, NC; Kristy Cash, Wilson Elementary, Coppell, TX; Monica Coles, Swift Creek Elementary School, Raleigh, NC; Shirley Conner, Ocean Avenue Elementary School, Portland, ME; Sally Connolly, Cape Elizabeth Middle School, Cape Elizabeth, ME; Melissa Cook-Airhart, Harry McKillop Elementary, Melissa, TX; Melissa Costa, Olive Chapel Elementary, Apex, NC; Hillary P. Croissant, Harry McKillop Elementary, Melissa, TX; Rene Custeau, Hall Elementary School, Portland, ME; Nancy Davis, Martha and Josh Morriss Mathematics and Engineering Elementary School, Texarkana, TX; Nancy Deveneau, Wilson Elementary, Coppell, TX; Karen Diaz, Las Colinas Elementary, Irving, TX; Marlana Dumas, Las Colinas Elementary, Irving, TX; Mary Evans, R.E. Good Elementary School, Carrollton, TX; Jacquelyn Farley, Moss Haven Elementary, Dallas, TX; Corinna Ferrier, Oak Forest Elementary, Humble, TX; Allison Fike, Wilson Elementary, Coppell, TX; Barbara Fugitt, Martha and Josh Morriss Mathematics and Engineering Elementary School, Texarkana, TX; Colleen Garvey, Farmington Woods Elementary, Cary, NC; Judy Geller, Bentley Elementary School, Oakland, CA; Erin Gibson, Las Colinas Elementary, Irving, TX; Kelli Gobel, Melissa Ridge Intermediate School, Melissa, TX; Dollie Green, Melissa Ridge Intermediate School, Melissa, TX; Brenda Lee Harrigan, Bentley Elementary School, Oakland, CA; Cori Harris, Samuel Beck Elementary, Trophy Club, TX; Kim Hayes, Martha and Josh Morriss Mathematics and Engineering Elementary School, Texarkana, TX; Staci Lynn Hester, Lacy Elementary School, Raleigh, NC; Amanda Hill, Las Colinas Elementary, Irving, TX; Margaret Hillman, Ocean Avenue Elementary School, Portland, ME; Cindy Holder, Oak Forest Elementary, Humble, TX; Sarah Huber, Hodge Road Elementary, Knightdale, NC; Susan Jacobs, Granger Elementary, Keller, TX; Carol Kellum, Wallace Elementary, Dallas, TX; Jennifer A. Kelly, Hall Elementary School, Portland, ME; Brittani Kern, Fox Road Elementary, Raleigh, NC; Jodi Lay, Lufkin Road Middle School, Apex, NC; Melissa Lourenco, Lake Myra Elementary School, Wendell, NC; Ana Martinez, RISD Academy, Dallas, TX; Shaheen Mavani, Las Colinas Elementary, Irving, TX; Mary Linley McClendon, Math Science Technology Magnet School, Richardson, TX; Adam McKay, Davis Drive Elementary, Cary, NC; Leslie Meadows, Lake Myra Elementary School, Wendell, NC; Anne Mechler, J. Erik Jonsson Community School, Dallas, TX; Anne Miller, J. Erik Jonsson Community School, Dallas, TX; Shirley Diann Miller, The Rice School, Houston, TX; Keri Minier, Las Colinas Elementary, Irving, TX; Stephanie Renee Nance, T.H. Rogers Elementary, Houston, TX; Cynthia Nilsen, Peaks Island School, Peaks Island, ME; Elizabeth Noble, Las Colinas Elementary, Irving, TX; Courtney Noonan, Shadow Oaks Elementary School, Houston, TX; Sarah Peden, Aversboro Elementary School, Garner, NC; Carrie Prince, School at St. George Place, Houston, TX; Marlaina Pritchard, Melissa Ridge Intermediate School, Melissa, TX; Alice Pujol, J. Erik Jonsson Community School, Dallas, TX; Claire Ramsbotham, Cape Elizabeth Middle School, Cape Elizabeth, ME; Paul Rendon, Bentley Elementary, Oakland, CA; Janette Ridley, W.H. Wilson Elementary School, Coppell, TX; Kristina (Crickett) Roberts, W.H. Wilson Elementary School, Coppell, TX; Heather Rogers, Wendell Creative Arts & Science Magnet Elementary School, Wendell, NC; Alissa Royal, Melissa Ridge Intermediate School, Melissa, TX; Megan Runion, Olive Chapel Elementary, Apex, NC; Christy Scheef, J. Erik Jonsson Community School, Dallas, TX; Samrawit Shawl, T.H. Rogers School, Houston, TX; Nicole Spivey, Lake Myra Elementary School, Wendell, NC; Ashley Stephenson, J. Erik Jonsson Community School, Dallas, TX; Jolanta Stern, Browning Elementary School, Houston, TX; Gale Stimson, Bentley Elementary, Oakland, CA; Ted Stoeckley, Hall Middle School, Larkspur, CA; Cathryn Sutton, Wilson Elementary, Coppell, TX; Camille Swander, Ocean Avenue Elementary School, Portland, ME; Brandi Swann, Westlawn Elementary School, Texarkana, TX; Robin Taylor, Arapaho Classical Magnet, Richardson, TX; Michael C. Thomas, Forest Lane Academy, Dallas, TX; Jomarga Thompkins, Lockhart Elementary, Houston, TX; Mary Timar, Madera Elementary, Lake Forest, CA; Helena Tongkeamha, White Rock Elementary, Dallas, TX; Linda Trampe, J. Erik Jonsson Community School, Dallas, TX; Charity VanHorn, Fred A. Olds Elementary, Raleigh, NC; Kathleen VanKeuren, Lufkin Road Middle School, Apex, NC; Valerie Vassar, Hall Elementary School, Portland, ME; Megan Veron, Westwood Elementary School, Houston, TX; Mary Margaret Waters, Frances E. Norton Elementary, Allen, TX; Stephanie Robledo Watson, Ridgecrest Elementary School, Houston, TX; Lisa Webb, Madisonville Intermediate, Madisonville, TX; Matt Whaley, Cape Elizabeth Middle School, Cape Elizabeth, ME; Nancy White, Canyon Creek Elementary, Austin, TX; Barbara Yurick, Oak Forest Elementary, Humble, TX; Linda Zittel, Mira Vista Elementary, Richmond, CA

**Photo Credits:** © iStockphoto/Ken Canning (cover); © Zach Smith; © Monkey Business Images/Shutterstock; © John Quick; © Christian Musat/Shutterstock

**Published and Distributed by Delta Education, a member of the School Specialty Family**
The FOSS program was developed in part with the support of the National Science Foundation grant nos. MDR-8751727 and MDR-9150097. However, any opinions, findings, conclusions, statements, and recommendations expressed herein are those of the authors and do not necessarily reflect the views of NSF. FOSSmap was developed in collaboration between the BEAR Center at UC Berkeley and FOSS at the Lawrence Hall of Science.

Environments — Teacher Toolkit, 1487682
Teacher Resources, 1487596
978-1-62571-334-6
Printing 4 – 6/2017
Patterson Printing, Benton Harbor, MI

This warning label is required by the
U.S. Consumer Product Safety Commission.
The chemical in the FOSS Environments kit is
kosher salt.

## TABLE OF CONTENTS

FOSS Program Goals

Science Notebooks in Grades 3–5

Science-Centered Language Development

FOSS and Common Core ELA — Grade 4

FOSS and Common Core Math — Grade 4

Taking FOSS Outdoors

Science Notebook Masters

Teacher Masters

Assessment Masters

This document, *Teacher Resources*, is one of three parts of the *FOSS Teacher Toolkit* for this module. The chapters in *Teacher Resources* are all available as PDFs on FOSSweb.

The other parts of the module *Teacher Toolkit* are the *Investigations Guide* and a copy of the *FOSS Science Resources* student book containing original readings for this module.

The spiral-bound *Investigations Guide* contains these chapters.

• Overview
• Framework and NGSS
• Materials
• Technology
• Investigations
• Assessment

The *Teacher Toolkit* is the most important part of the FOSS Program. It is here that all the wisdom and experience contributed by hundreds of educators has been assembled. Everything we know about the content of the module, how to teach the subject, and the resources that will assist the effort are presented here.

# FOSS Program
# Goals

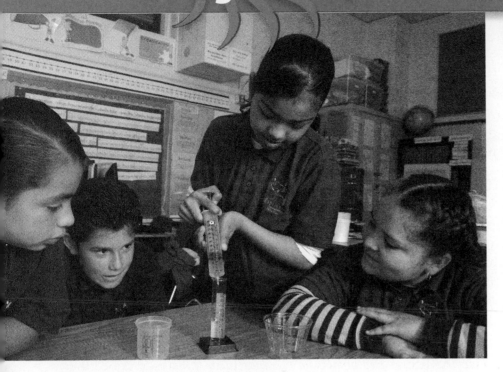

## Contents

Introduction ............................1

Goals of the FOSS Program .....2

Bridging Research
into Practice ............................5

FOSS Next Generation K–8
Scope and Sequence ................8

# INTRODUCTION

*The Full Option Science System™ has evolved from a philosophy of teaching and learning at the Lawrence Hall of Science that has guided the development of successful active-learning science curricula for more than 40 years. The FOSS Program bridges research and practice by providing tools and strategies to engage students and teachers in enduring experiences that lead to deeper understanding of the natural and designed worlds.*

Science is a creative and analytic enterprise, made active by our human capacity to think. Scientific knowledge advances when scientists observe objects and events, think about how they relate to what is known, test their ideas in logical ways, and generate explanations that integrate the new information into understanding of the natural world. Engineers apply that understanding to solve real-world problems. Thus, the scientific enterprise is both what we know (content knowledge) and how we come to know it (practices). Science is a discovery activity, a process for producing new knowledge.

The best way for students to appreciate the scientific enterprise, learn important scientific and engineering concepts, and develop the ability to think well is to actively participate in scientific practices through their own investigations and analyses. FOSS was created to engage students and teachers with meaningful experiences in the natural and designed worlds.

# GOALS OF THE FOSS PROGRAM

FOSS has set out to achieve three important goals: scientific literacy, instructional efficiency, and systemic reform.

## Scientific Literacy

FOSS provides all students with science experiences that are appropriate to students' cognitive development and prior experiences. It provides a foundation for more advanced understanding of core science ideas which are organized in thoughtfully designed learning progressions and prepares students for life in an increasingly complex scientific and technological world.

The National Research Council (NRC) in *A Framework for K–12 Science Education* and the American Association for the Advancement of Science (AAAS) in *Benchmarks for Scientific Literacy,* have described the characteristics of scientific literacy:

- Familiarity with the natural world, its diversity, and its interdependence.

- Understanding the disciplinary core ideas and the cross-cutting concepts of science, such as patterns; cause and effect; scale, proportion, and quantity; systems and system models; energy and matter—flows, cycles, and conservation; structure and function; and stability and change.

- Knowing that science and engineering, technology, and mathematics are interdependent human enterprises and, as such, have implied strengths and limitations.

- Ability to reason scientifically.

- Using scientific knowledge and scientific and engineering practices for personal and social purposes.

The FOSS Program design is based on learning progressions that provide students with opportunities to investigate core ideas in science in increasingly complex ways over time. FOSS starts with the intuitive ideas that primary students bring with them and provides experiences that allow students to develop more sophisticated understanding as they grow through the grades. Cognitive research tells us that learning involves individuals in actively constructing schemata to organize new information and to relate and incorporate the new understanding into established knowledge. What sets experts apart from novices is that experts in a discipline have extensive knowledge that is effectively organized into structured schemata to promote thinking. Novices have disconnected ideas about a topic that are difficult to retrieve and use. Through internal processes to establish schemata and through social processes of interacting with peers and adults, students construct understanding of the natural world and their relationship to it. The target goal for FOSS students is to know and use scientific explanations of the natural world and the designed world; to understand the nature and development of scientific knowledge and technological capabilities; and to participate productively in scientific and engineering practices.

## Instructional Efficiency

FOSS provides all teachers with a complete, cohesive, flexible, easy-to-use science program that reflects current research on teaching and learning, including student discourse, argumentation, writing to learn, and reflective thinking, as well as teacher use of formative assessment to guide instruction. The FOSS Program uses effective instructional methodologies, including active learning, scientific practices, focus questions to guide inquiry, working in collaborative groups, multisensory strategies, integration of literacy, appropriate use of digital technologies, and making connections to students' lives, including the outdoors.

FOSS is designed to make active learning in science engaging for teachers as well as for students. It includes these supports for teachers:

- Complete equipment kits with durable, well-designed materials for all students.

- Detailed *Investigations Guide* with science background for the teacher and focus questions to guide instructional practice and student thinking.

- Multiple strategies for formative assessment at all grade levels.

- Benchmark assessments (grades 1–5) with online access for administering, coding, and analyzing assessments (grades 3–5).

- Strategies for use of science notebooks for novice and experienced users.

- *FOSS Science Resources,* a book of module-specific readings with strategies for science-centered language development.

- The FOSS website with interactive multimedia activities for use in school or at home, suggested interdisciplinary-extension activities, and extensive online support for teachers, including teacher prep videos.

## Systemic Reform

FOSS provides schools and school systems with a program that addresses the community science-achievement standards. The FOSS Program prepares students by helping them acquire the knowledge and thinking capacity appropriate for world citizens.

The FOSS Program design makes it appropriate for reform efforts on all scales. It reflects the core ideas to be incorporated into the next-generation science standards. It meets with the approval of science and technology companies working in collaboration with school systems, and it has demonstrated its effectiveness with diverse student and teacher populations in major urban reform efforts. The use of science notebooks and formative-assessment strategies in FOSS redefines the role of science in a school—the way that teachers engage in science teaching with one another as professionals and with students as learners, and the way that students engage in science learning with the teacher and with one another. FOSS takes students and teachers beyond the classroom walls to establish larger communities of learners.

# BRIDGING RESEARCH INTO PRACTICE

The FOSS Program is built on the assumptions that understanding core scientific knowledge and how science functions is essential for citizenship, that all teachers can teach science, and that all students can learn science. The guiding principles of the FOSS design, described below, are derived from research and confirmed through FOSS developers' extensive experience with teachers and students in typical American classrooms.

***Understanding of science develops over time.*** FOSS has elaborated learning progressions for core ideas in science for kindergarten through grade 6. Developing the learning progressions involves identifying successively more sophisticated ways of thinking about core ideas over multiple years. "If mastery of a core idea in a science discipline is the ultimate educational destination, then well-designed learning progressions provide a map of the routes that can be taken to reach that destination" (National Research Council, *A Framework for K–12 Science Education*, 2012).

***Focusing on a limited number of topics in science avoids shallow coverage and provides more time to explore core science ideas in depth.*** Research emphasizes that fewer topics experienced in greater depth produces much better learning than many topics briefly visited. FOSS affirms this research. FOSS modules provide long-term engagement (9–10 weeks) with important science ideas. Furthermore, modules build upon one another within and across each strand, progressively moving students toward the grand ideas of science. The core ideas of science are difficult and complex, never learned in one lesson or in one class year.

### FOSS Next Generation—Elementary Module Sequences

| | PHYSICAL SCIENCE | | EARTH SCIENCE | | LIFE SCIENCE | |
|---|---|---|---|---|---|---|
| | MATTER | ENERGY AND CHANGE | DYNAMIC ATMOSPHERE | ROCKS AND LANDFORMS | STRUCTURE/ FUNCTION | COMPLEX SYSTEMS |
| 5 | Mixtures and Solutions | | Earth and Sun | | Living Systems | |
| 4 | | Energy | | Soils, Rocks, and Landforms | Environments | |
| 3 | Motion and Matter | | Water and Climate | | Structures of Life | |
| 2 | Solids and Liquids | | | Pebbles, Sand, and Silt | Insects and Plants | |
| 1 | | Sound and Light | Air and Weather | | Plants and Animals | |
| K | Materials and Motion | | Trees and Weather | | Animals Two by Two | |

**Science is more than a body of knowledge.** How well you think is often more important than how much you know. In addition to the science content framework, every FOSS module provides opportunities for students to engage in and understand science practices, and many modules explore issues related to engineering practices and the use of natural resources. FOSS promotes these science and engineering practices described in *A Framework for K–12 Science Education.*

- Asking questions (for science) and defining problems (for engineering)

- Developing and using models

- Planning and carrying out investigations

- Analyzing and interpreting data

- Using mathematics and computational thinking

- Constructing explanations (for science) and designing solutions (for engineering)

- Engaging in argument from evidence

- Obtaining, evaluating, and communicating information

**Science is inherently interesting, and children are natural investigators.** It is widely accepted that children learn science concepts best by doing science. Doing science means hands-on experiences with objects, organisms, and systems. Hands-on activities are motivating for students, and they stimulate inquiry and curiosity. For these reasons, FOSS is committed to providing the best possible materials and the most effective procedures for deeply engaging students with scientific concepts. FOSS students at all grade levels investigate, experiment, gather data, organize results, and draw conclusions based on their own actions. The information gathered in such activities enhances the development of scientific and engineering practices.

**Education is an adventure in self-discovery.** Science provides the opportunity to connect to students' interests and experiences. Prior experiences and individual learning styles are important considerations for developing understanding. Observing is often equated with seeing, but in the FOSS Program all senses are used to promote greater understanding. FOSS evolved from pioneering work done in the 1970s with students with disabilities. The legacy of that work is that FOSS investigations naturally use multisensory methods to accommodate students with physical and learning disabilities and also to maximize information gathering for all students. A number of tools, such as the FOSS syringe and balance, were originally designed to serve the needs of students with disabilities.

**Formative assessment is a powerful tool to promote learning and can change the culture of the learning environment.** Formative assessment in FOSS creates a community of reflective practice. Teachers and students make up the community and establish norms of mutual support, trust, respect, and collaboration. The goal of the community is that everyone will demonstrate progress and will learn and grow.

**Science-centered language development promotes learning in all areas.** Effective use of science notebooks can promote reflective thinking and contribute to life long learning. Research has shown that when language-arts experiences are embedded within the context of learning science, students improve in their ability to use their language skills. Students are eager to read to find out information, and to share their experiences both verbally and in writing.

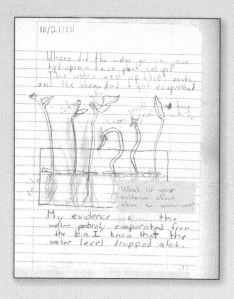

**Experiences out of the classroom develop awareness of community.** By extending classroom learning into the outdoors, FOSS brings the science concepts and principles to life. In the process of validating classroom learning among the schoolyard trees and shrubs, down in the weeds on the asphalt, and in the sky overhead, students will develop a relationship with nature. It is our relationship with natural systems that allows us to care deeply for these systems.

# FOSS Program Goals

## FOSS NEXT GENERATION K–8 SCOPE AND SEQUENCE

| Grade | Physical Science | Earth Science | Life Science |
|---|---|---|---|
| 6–8 | Waves*<br>Gravity and Kinetic Energy* | Planetary Science | Heredity and Adaptation*<br>Human Systems Interactions* |
| | Chemical Interactions | Earth History | Populations and Ecosystems |
| | Electromagnetic Force*<br>Variables and Design* | Weather and Water | Diversity of Life |
| 5 | Mixtures and Solutions | Earth and Sun | Living Systems |
| 4 | Energy | Soils, Rocks, and Landforms | Environments |
| 3 | Motion and Matter | Water and Climate | Structures of Life |
| 2 | Solids and Liquids | Pebbles, Sand, and Silt | Insects and Plants |
| 1 | Sound and Light | Air and Weather | Plants and Animals |
| K | Materials and Motion | Trees and Weather | Animals Two by Two |

\* Half-length course

FOSS is a research-based science curriculum for grades K–8 developed at the Lawrence Hall of Science, University of California, Berkeley. FOSS is also an ongoing research project dedicated to improving the learning and teaching of science. The FOSS Program materials are designed to meet the challenge of providing meaningful science education for all students in diverse American classrooms and to prepare them for life in the 21st century. Development of the FOSS Program was, and continues to be, guided by advances in the understanding of how people think and learn.

With the initial support of the National Science Foundation and continued support from the University of California, Berkeley, and School Specialty, Inc., the FOSS Program has evolved into a curriculum for all students and their teachers, grades K–8.

# Science Notebooks
# in Grades 3-5

*Source: Special Collections Research Center, University of Chicago Library*

Enrico Fermi's laboratory notebook, 1941

Student notebook from the Energy and Electromagnetism Module

# INTRODUCTION

A scientist's notebook is a detailed record of his or her engagement with natural phenomena. It is a personal representation of experiences, observations, and thinking—an integral part of the process of doing scientific work. A scientist's notebook is a continuously updated history of the development of scientific knowledge and reasoning. FOSS students are young scientists; they incorporate notebooks into their science learning.

This chapter is designed to be a resource for teachers who are incorporating notebooks into their classroom practice. For teachers just beginning to use notebooks, the Getting Started section in this chapter suggests how to set up the notebooks, and the *Investigations Guide* cues you when to engage students with the notebooks during the investigation. For more information on specific types of notebook entries, the subsections in the Notebook Components section include strategies to differentiate instruction for various ability levels.

## Contents

Introduction ............................1
Notebook Benefits ....................2
Getting Started ........................7
Notebook Components ..........12
  Planning the Investigation .... 12
  Data Acquisition and Organization ......................15
  Making Sense of Data ......... 18
  Next-Step Strategies............22
Writing Outdoors...................26
Closing Thoughts..................28

**Full Option Science System**     ***Copyright © The Regents of the University of California***

# NOTEBOOK BENEFITS

Engaging in active science is one part experience and two parts making sense of the experience. Science notebooks help students with the sense-making part. Science notebooks assist with documentation and cognitive engagement. For teachers, notebooks are tools for gaining insight into students' thinking. Notebooks inform and refine instructional practice.

## Benefits to Students

**Benefits to Students**
- *Documentation*
- *Reference document*
- *Cognitive engagement*

***Documentation.*** Science provides an authentic experience for students to develop their documentation skills. Students are encouraged to keep science notebooks to organize and learn to communicate their thinking. They document their experiences, data, and thinking during each investigation. Students create tables, graphs, charts, drawings, and labeled illustrations as standard means for representing and displaying data. At first, students will look at their science notebooks as little more than a random collection of words and pictures. Each notebook page represents an isolated activity. As students become more accomplished at keeping notebooks, their documentation will become better organized and efficient. In time and with some guidance, students will adopt a deeper understanding of their collections as integrated records of their learning.

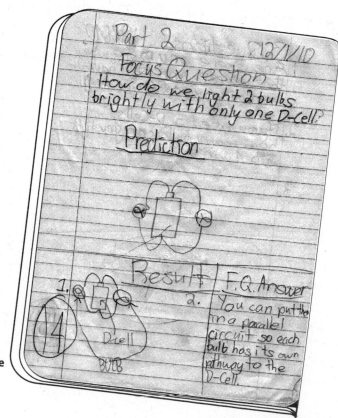

**Science notebook entry from the Energy and Electromagnetism Module**

**Reference document.** When data are displayed in functional ways, students can think about them more effectively. A well-kept notebook is a useful reference document. When students have forgotten a fact about a rock that they learned earlier in their studies, they can look it up. Learning to trust a personal record of previous discoveries and knowledge structures is important.

A complete and accurate record allows students to reconstruct the sequence of learning events to "relive" the experience. Discussions about science among students; students and teachers; or students, teachers, and families have more meaning when they are supported by authentic documentation in students' notebooks.

**From the Water and Landforms Module**

***Cognitive engagement.*** Once data are recorded and organized in an efficient manner in science notebooks, students can think about them to draw conclusions about the way the world works. Their data, based on their experiences and observations, are the raw materials that students use to forge concepts and relationships.

Writing stimulates active reasoning. There is a direct relationship between the formation of concepts and the rigors of expressing those thoughts in words. Writing requires students to impose discipline on their thoughts. By writing explanations, students clarify what they know . . . and expose what they don't know.

When students are asked to generate derivative products (such as detailed explanations or posters) as evidence of learning, the process will be much more efficient when they have a coherent, detailed notebook for reference. As students begin to use their notebooks as a personal reference text, they will value their own learning and come to rely on their own work as a source of information about science.

When students use their notebooks as an integral part of their science studies, they are more likely to think critically about their understandings. This reflective thinking can be encouraged by notebook entries that present opportunities for self-assessment. Self-assessment motivates students to rethink and restate their scientific understandings. Revising their notebook entries by using a next-step strategy helps students clarify their understanding of the science concepts under investigation.

▶ **NOTE**
For more on derivative products, see the Science-Centered Language Development chapter.

## Benefits to Teachers

In FOSS, the unit of instruction is the module—a sequence of conceptually related learning experiences that leads to a set of learning outcomes. A science notebook helps you think about and communicate the conceptual structure of the module you are teaching.

**Assessment.** From the assessment point of view, a science notebook is a collection of student-generated artifacts that exhibit learning. You can assess student skills, such as using drawings to record data, while students are working with materials. At other times, you collect the notebooks and review them in greater detail. The displays of data and analytical work, such as responses to focus questions, provide a measure of the quality and quantity of student learning. The notebook itself should not be graded. However, the notebook can be considered as one component of a student's overall performance in science.

**Medium for feedback.** The science notebook is an excellent medium for providing feedback to individual students regarding their work. Most students will be able to read a teacher comment written on a self-stick note, think about the issue, and respond. Some students may need oral feedback individually or in a small-group situation. This feedback might include modeling to help students make more accurate drawings, revisiting some essential scientific vocabulary, or introducing strategies to better explain their thinking.

**Focus for professional discussions.** The science notebook acts as a focal point for discussion about students' learning at several levels. It can be reviewed and discussed during parent conferences. Science notebooks can be the focus of three-way discussions among students, teachers, and principals to ensure that all members of the school science community agree about what kinds of student work are valued and the level of performance to expect. Science notebooks shared among teachers in a study group or other professional-development environment can serve as a reflective tool that informs teachers of students' ability to demonstrate recording techniques, individual styles, various levels of good-quality work, and so on. Just as students can learn notebook strategies from one another, teachers can learn notebook skills from one another.

**Benefits to Teachers**
- *Assessment*
- *Medium for feedback*
- *Focus for professional discussions*
- *Refinement of practice*

**Refinement of practice.** As teachers, we are constantly looking for ways to improve instructional practices to increase students' understanding. Your use of the notebook should change over time. In the beginning, the focus will be on the notebook itself—what it looks like, what goes in it. As you become more comfortable with the notebook, the attention shifts to what students are learning. When this happens, you begin to consider how much scaffolding to provide to different students, how to use evidence of learning to differentiate instruction, and how to modify instruction to refine students' understanding.

# GETTING STARTED

A major goal for using notebooks is to establish habits that will enable students to collect data and make sense of them. Use of the notebook must be flexible enough to allow students room to grow and supportive enough for students to be successful from the start. The format should be simple and the information meaningful to students. The notebook includes student drawings; writing in the form of individual words, short phrases, and sentences; and a variety of visual and tactile artifacts. When students thumb through their notebooks, they should be reminded of the objects and organisms they observed and their interactions with them.

## Notebook Format

The *Teacher Resources* component of the FOSS *Teacher Toolkit* includes duplication masters for the same notebook sheets that are bound into the consumable notebooks. You can use these sheets to prepare an analogous notebook or to have students design a customized version in a composition book.

In an autonomous approach, students create their entire science notebooks from blank pages in bound composition books. Experienced students determine when to use their notebooks, how to organize space, what methods of documentation to use, and how to flag important information. This level of notebook use will not be realized quickly; it will likely require systematic development by an entire teaching staff over several years.

You might choose to have a separate notebook for each module or one notebook for the entire year. (See the sidebar for the advantages of each.) Students will need about 30 pages (60 sides) for a typical module.

**Advantages of One Notebook per Module**
- *Easy to replace if lost*
- *Lower cost*
- *Fewer pages*

**Advantages of One Notebook per Year**
- *Easy to refer to prior activities*
- *Easy to see growth over time*

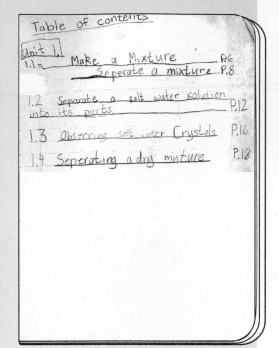

**A sample table of contents**

## Organization of Notebooks

Four organizational components of the notebook should be planned right from the outset: a table of contents, page numbering, documentation, and an index. The consumable FOSS science notebook has these features already in place. No setup is required beyond pointing out where these features are in the notebook.

*Table of contents.* Students should reserve the first two pages of their notebooks for a table of contents. They will add to it systematically as they proceed through the module. The table of contents can be based on the names of the investigations in the module, the specific activities undertaken, the concepts learned, or some other schema that makes sense to everyone. You could have students make their own titles for the table of contents. This is best done at the end of the investigation part so that students can come up with meaningful titles.

*Page numbering.* Each page should have a number. These can be applied to the pages and referenced in the table of contents as the notebook progresses, or small blocks of pages can be prenumbered (pages 1–15 initially, pages 16–30 later, and so on) at appropriate times in the module.

*Date, title, and other conditions.* Each time students make a new entry, they should record certain information. At the very minimum, they should record the date and a title. More complete documentation might include the time; day of the week; team members; and, if appropriate, weather conditions.

Some classes start each new entry at the top of the next available page. Others simply leave a modest space and apply the documentation information right before the new entry.

When introducing a new condition to students, such as weather conditions, it is important to discuss why students are recording the information so that they understand its relevancy.

**Index.** Scientific academic language is important. FOSS strives to have students use precise, accurate vocabulary at all times in their writing and conversations. To assist with acquisition of the scientific vocabulary, students should set up an index at the end of their notebooks. It is not usually possible to enter the words in alphabetical order, since they will be acquired as the module advances. Instead, assign a block of letters to each of several index pages (A–E, F–K, etc.) where students can enter key words. As an alternative, students could use a single page blocked out in 24 squares and assign one or more letters to each square. (See the sample illustration.) Students write the new vocabulary word or phrase in the appropriate square and tag it with the number of the page on which the word is defined in the notebook.

## Notebook Entries

As students engage in scientific exploration, they will make entries in their notebooks. They may use a prepared notebook sheet or create a more free-form entry. Students frequently respond to a focus question with a drawing or a simple narrative entry. For example, in the **Soils, Rocks, and Landforms Module**, students may explain their thinking about factors that affect erosion and include a drawing to show how the earth materials move when the land is steep. These notebook entries allow students to relive and describe their science experiences as they turn the pages in their notebooks.

Typically, the rules of grammar and spelling are relaxed when making notebook entries so as not to inhibit the flow of creative expression. Encourage students to use many means of recording and communicating besides writing, including charts, graphs, drawings, graphics, color codes, numbers, and images attached to the notebook pages. By exploring many options for making notebook entries, each student will find his or her most efficient, expressive way to capture and organize information for later retrieval.

▶ **NOTE**
Templates for indexes can be found online at www.FOSSweb.com.

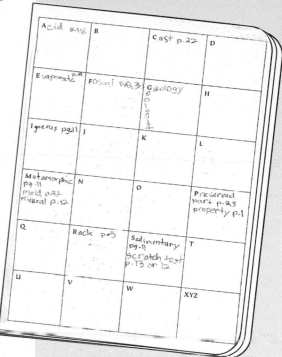

**A sample index**

▶ **NOTE**
More information about vocabulary appears in the Science-Centered Language Development chapter.

## Supporting Students

Elementary classrooms contain students with a range of abilities, which need to be considered when planning strategies for implementing science notebooks. Students need to have successful early experiences with notebooks. A blank notebook may be intimidating for some students, and they will look to you for guidance. FOSS teachers have had success using different supportive structures to help transform the blank notebook into a valuable reference tool. These supports and scaffolds can be used with the whole class, a small group, or an individual and should be adjusted to meet students' needs.

*Class notebook.* You can create a class notebook to document the investigation as a way to model the various notebook components. The class notebook should be accessible at all times for students to reference. You can use a chart-paper tablet, a paper notebook displayed under a document camera, or a computer. You could use the class notebook to introduce strategies such as a T-table, or write a summary statement in it after all students have answered the focus question. While individual notebooks will look similar to the class notebook, it is not the intent that students' notebooks be identical to the class notebook.

*Scaffolds.* Supports and scaffolds differ in one way. Supports are always available for students to access, such as allowing students access to a class notebook. Scaffolds are available just when the student needs them and will vary from student to student and from investigation to investigation. Scaffolds are meant to provide structure to a notebook entry and allow students to insert their observations or thinking. As the year progresses, the scaffolds change to allow for more student independence. Scaffolds include

- **Sentence starters** or **drawing starters** provide a beginning point for a notebook entry. Here's an example: "I noticed the crayfish had _____."

- **Frames** provide more support but leave specific gaps for students to complete. Here's an example: "Magnets attract when _____. Magnets repel when _____."

- The suggested **notebook sheets** can guide students to collect data with a table, graph, or list of questions. The notebook sheets also guide thinking.

**TEACHING NOTE**

*Use of the class notebook should be thoughtfully timed. Doing a class-notebook entry at the end of an activity is helpful to teach the components of a notebook, yet allows you to see what students do on their own. If you want to model a specific notebook strategy, use the class notebook during the activity.*

***Think-alouds.*** Think-alouds help explain the decision-making process practiced by a savvy notebook user. They verbalize the thoughts used to create a particular notebook entry. For example, if students have recorded observations about how one variable affects erosion and deposition and are going to observe a second variable, you might say,

*I am going to observe how another variable affects erosion and deposition. I am going to look back to see how I recorded the information before. I see that I made a detailed drawing. I used a T-table to describe what happened to the earth materials over time. So now I'm going to make similar observations and use a T-table and then make a drawing. One way I can get ideas about how to organize my observations is to look back at observations I wrote before.*

***Providing time to record.*** When students are engaged in active science, their efforts are focused on the materials, not the notebook. Students need this time to explore and initially might not open their notebooks and record observations until prompted. Students may need separate time to record observations that fully document their discoveries. Some teachers have found it easier to leave the materials on the table and have students bring their notebooks to a common writing area. Then the teacher revisits the focus question or task and provides a few minutes for students to record in their notebooks.

## Ownership

A student's science notebook can be personal or public. If the notebook is personal, the student decides how accessible his or her work is to other students. If ownership falls at the opposite extreme, everything is public, and anyone can look at the contents of anyone else's notebook at any time. In practice, most classroom cultures establish a middle ground in which a student's notebook is substantially personal, but the teacher claims free access to the students' work and can request that students share notebooks with one another and with the whole class from time to time.

**Notebook Components**

- *Planning the investigation*
- *Data acquisition and organization*
- *Making sense of data*
- *Next-step strategies*

# NOTEBOOK COMPONENTS

A few components give the science notebook conceptual shape, coherence, and direction. These components don't prescribe a step-by-step procedure for how to prepare the notebook, but they do provide some overall guidance.

The general arc of an investigation starts with a question or challenge, and then proceeds with an activity, data acquisition, sense making, and next steps. The science notebook should record important observations and thoughts along the way. It may be useful to keep these four components in mind as you systematically guide students through their notebook entries.

## Planning the Investigation

Typically at the start of a new activity, the first notebook entry is a focus question, which students transcribe into their notebooks. The focus question establishes the direction and conceptual challenge for the investigation.

***Focus question.*** Each part of each investigation starts with a focus question or challenge. Write or project it on the board for students to transcribe into their notebooks, or give them photocopied strips of the focus question to tape or glue into their notebooks. Some teachers look ahead, write all the focus questions on one sheet of paper, copy the sheet, and cut the questions apart with a paper cutter. A list of the focus questions for each module is also available as a PDF on FOSSweb.

> 2. **Focus question: What's in our schoolyard soils?**
> Write or project the focus question on the board, and say it aloud.
>
> ➤ *What's in our schoolyard soils?*

Students may develop their own questions to investigate as well.

The focus question determines the kinds of data to be collected and the procedures that will yield those data. The procedures may be a narrative plan consisting of a few sentences or a more detailed step-by-step procedure, depending on the requirements of the investigation.

**Narrative plans.** After posing a focus question, we often ask students for their ideas about how they will engage the question. For instance, the focus question that sets up a free exploration of crayfish might be

➤ *What do crayfish do when they are removed from water for a short time?*

Students think about this question and formulate a plan, perhaps writing a short narrative description of their general approach.

*We are going to take the crayfish out and put it on the table for 3 minutes. We are going to see if the swimmerets move the same way they do in the water.*

The narrative helps students consider the options available to them. It also reminds them of the limits and considerations when working with living organisms.

As with all notebook entries, some students need little more than a nudge in the right direction. Other students may need more support. One option is to prompt them with sentence starters. Good sentence starters provide a start but still leave the intellectual responsibility with students. Here are some examples.

- *The first thing we are going to do is _____.*
- *The next thing we will do is _____.*
- *We have to be careful about _____.*
- *Finally, we are going to _____.*

**Lists.** Science notebooks often include lists of things to think about, materials to get, or words to remember. A materials list is a good organizer, helping students anticipate actions they will take. A list of variables to be controlled clarifies the purpose of an experiment. Simple lists of dates for making observations, or of the people responsible for completing a task keep information readily available.

> **TEACHING NOTE**
>
> *Supports such as sentence starters and frames should be monitored and adjusted as your expectation of students' responsibility for notebooks changes over the year.*

**TEACHING NOTE**

*Be selective when engaging students in making step-by-step procedures as the process will require additional time. To save time when using a class-generated procedure, provide each student with a typed copy.*

**TEACHING NOTE**

*It is important for students to understand that in order to do accurate work, scientists need to do multiple trials and present their work so that it can be replicated.*

**Step-by-step procedures.** Some work with materials requires structured planning. Students start an investigation in the **Soils, Rocks, and Landforms Module** with the focus question

➤ *What's in our schoolyard soils?*

Students need to recall what they know about soil and develop a strategy for adding water to help isolate the different components. An appropriate convention for recording a sequential procedure is a numbered, step-by-step plan. Here is an example.

1. Collect 1 spoon of soil.

2. Record where it came from.

3. Put it in a vial.

4. Add 20 mL of water.

5. Cap the vial and shake.

6. Let it settle overnight.

7. Observe layers.

One way to introduce students to this type of entry is to provide a notebook sheet as a model. Both you and students can refer to the notebook sheet as they work through the hands-on investigation. During the next investigation, students can look back at the model notebook sheet when they write their own step-by-step procedures. To check the procedures for errors or omissions, students can trade notebooks and attempt to follow other students' instructions to complete the task.

**Predictions.** Depending on the content and the focus question, students may be able to make a prediction. When they make predictions, they are attempting to relate prior experiences to the question posed. Providing students with a frame can help them explain the rationale behind their predictions. A frame to help with stating a prediction is "I think that _____ because _____."

## Data Acquisition and Organization

Data are the bits of information (observations) from which scientists construct ideas about the structure and behaviors of the natural world. Because observation is the starting point for answering the focus question, data records should be

- clearly related to the focus question;

- accurate and precise, including units with measurements;

- organized for efficient reference.

Data handling can have two phases: data acquisition and data display. Data acquisition is making observations and recording data. The data record can be composed of words, phrases, numbers, and/or drawings. Data display is reorganizing the data in a logical way to facilitate thinking. The display can take the form of narratives, drawings, artifacts, charts, tables, images, graphs, Venn diagrams, calendars, or other graphic organizers. Early in a student's experience with notebooks, the record may be disorganized and incomplete, and the display may need guidance. With practice, however, students will become skilled at determining what form of recording to use in various situations and how best to display the data for analysis.

***Narratives.*** The most intuitive approach to recording data for most students is narrative—using words, sentence fragments, and numbers in a more or less sequential manner. As students make a new observation, they record it right after the previous entry, followed by the next observation, and so on. Some observations, such as a record of the movements and interactions of a crayfish over time or the changes observed in an environment, are appropriately recorded in narrative form.

Sentence frames can help students record their narratives initially. Another option is to have students read their narratives to a partner from a different group. This allows them to refine their language and gain insight into another student's observation.

**A drawing from the Structures of Life Module**

**Drawings.** When students observe the shape and distribution of crystals after evaporation, or observe and identify the parts of a system, a labeled illustration is the most efficient way to record data. A picture is worth a thousand words, and a labeled picture is even more useful.

Students' initial drawings may need refinement to accurately capture their observations. It can be helpful to suggest an acronym for making useful drawings. Accurate, big, colorful, and detailed (ABCD) drawings can explain systems, identify structures of an organism, or capture observations of landforms. As the need arises, introduce specific drawing techniques, such as the use of scale, magnified view, and appropriate use of color.

**Tables.** With experience, students will recognize when a table is appropriate for recording data. When students make similar observations about a series of objects, such as a set of minerals or powdered substances, a table with columns is an efficient recording method. The two-dimensional table makes it easy to compare the properties of all the objects under investigation. Similarly, when students conduct an experiment, they can record data directly into a T-table. With little effort, they can transform the table, presented in ordered pairs, into a graph.

Notebook sheets are a good way to scaffold the use of tables. Students can quickly enter the information on the sheets. As they take on more independence in their notebooks, discussions about the column headings or the purpose of the tables shift the focus from filling in the table, to the purpose the table serves, and, eventually, to how to make a table to record this information.

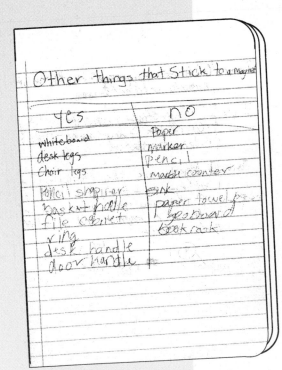

**A table from the Energy and Electromagnetism Module**

**Graphs, charts, and graphics.** Reorganizing data into a logical, easy-to-use graphic is the first phase of data analysis. Graphs allow easy comparison (bar graphs), quick statistical analysis of numerical data (line plots), and visual confirmation of a relationship between variables (two-coordinate graphs). Additional graphic tools, such as Venn diagrams, pie charts, concept maps, food chains, and life cycles, help students perceive patterns in their data.

As students progress in making graphs and graphics, so should their understanding of when and why to use them. Taking a little time to have students discuss which graph to use before distributing a notebook sheet will go a long way to reinforce the idea that different graphs serve different purposes.

**Artifacts.** Occasionally, the results of an investigation produce three-dimensional products that students can tape or glue directly into their science notebooks. Sand, minerals, seeds, and so on can become a permanent part of the record of learning.

**Images.** Digital photos of plants, rocks, and the results of investigations can be great additions to the science notebook. Digital photos should be used to enhance students' observations and drawings, not to replace them.

**A table from the Motion, Force, and Models Module**

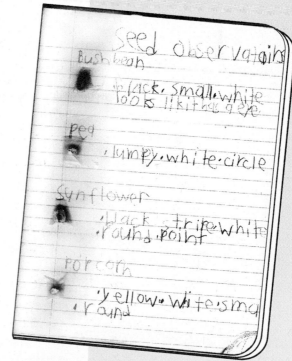

**Artifacts can be placed inside the notebook.**

## Making Sense of Data

Most of the sense making takes place during whole-group discussions when students share and discuss the observations they made while investigating. Questions are asked to help students interpret and analyze data in order to build conceptual understanding. Encourage students to use new vocabulary when they are sharing and making sense of their data. These words should be accessible to all students on the word wall. When students have limited written language skills, this oral discussion is important. This sharing is essential and is described in detail in the *Investigations Guide*. After sharing observations, the class revisits the focus question. Students flip back in their notebooks and use their data to discuss their answers to the question. You scaffold the discussion, and use appropriate language-development strategies.

The student's most important job is to think about the data and the focus question together to craft an answer to the question. Many investigations give students time to discuss and think with fellow students prior to writing an answer.

---

**14. Discuss the relative amounts of materials in soils**
Have students look at other groups' vials and the vials of soils you saved from Part 1 in order to compare the contents.

➤ *Do all vials contain the same amount of soil?* [Approximately.]

➤ *Do all vials have the same number of layers?* [Possibly, but not necessarily.]

➤ *Are the layers the same thickness for each type of earth material?* [Probably not.]

➤ *Can you identify each layer?* [Each soil should include some sand, a significant layer of silt and clay, and some humus.]

➤ *How are our schoolyard soils alike and different from the mountain, desert, river delta, and forest soils?*

---

After the sense-making discussion, students might review relevant vocabulary. They answer the focus question in their notebooks, expressing their current understanding. Some students will naturally use scientific vocabulary in their explanations, while others may need additional experiences before they include scientific vocabulary in their writing. In many instances, you can use supports to guide the development of a coherent and complete response to the question.

**Frames and prompts.** When providing frames or prompts, you are helping students organize their thinking. The frame provides a communication structure that allows students to focus their attention on thinking about the science involved. Here are some general frames.

- *I used to think _____, but now I think _____.*

- *The most important thing to remember about evaporation is _____.*

- *One thing I learned about _____.*

**TEACHING NOTE**

*Frames and prompts are designed to support students to either communicate or think more about the content. The important point is that students are doing the thinking and not merely responding to the prompt.*

When using frames, you need to provide just the right amount of scaffolding to allow students to communicate their understanding. Too much scaffolding limits the opportunity for students' communication skills to progress. Too little scaffolding may not allow the student to accurately represent his or her understanding.

For example, in the **Energy and Electromagnetism Module**, students are asked,

➤ *What is the relationship between the distance between two magnets and the force of attraction between them?*

You can provide different frames to students, depending on ability. Each frame has merit when used appropriately. The four different frames below give students progressively less support.

- *As the distance between the two magnets increases, the force of attraction_____. I know this because when I tested a distance of 6 spacers, the force broke with _____ washers. When I tested a distance of 1 spacer, the force broke with _____ washers.*

- *As the distance between the two magnets _____, the force of attraction _____. I know this because when I tested a distance of _____ spacers, the force broke with _____ washers. When I tested a distance of _____ spacer, the force broke with _____ washers.*

- *As the _____, the_____. I know this because _____.*

- *Make a claim about the force of attraction between magnets, and provide evidence to support your claim.*

**Claims and evidence.** A claim is an assertion about how the natural world works. A student might claim, for instance, that metals stick to magnets. For the claim to be valid and accurate, it must be supported by evidence—statements that are directly correlated with data. The evidence should refer to specific observations, relationships that are displayed in graphs, tables of data that show trends or patterns, dates, measurements, and so on. Many teachers will provide a frame to help students include both their claims and evidence in their responses. A claims–and–evidence construction is a sophisticated, rich display of student learning and thinking.

**Conclusions and predictions.** At the end of an investigation (major conceptual sequence), it may be appropriate for students to generate a summarizing narrative to succinctly communicate what they have learned. When appropriate, students can make predictions based on their understanding of a principle or relationship. For instance, after completing the investigation on evaporation, a student might predict the order in which the water will completely evaporate from various containers, based on surface area exposed to air. Or a student might predict how long to make a pendulum that swings 15 times in 15 seconds. Predictions will frequently indicate the degree to which a student can apply the new knowledge to real-world situations. A prediction can be the springboard for further inquiry by the class or by individual students.

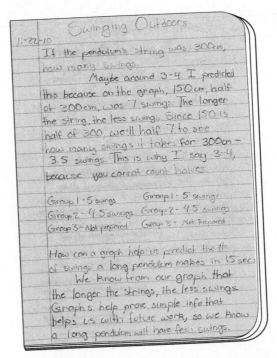

A prediction can show understanding of a relationship.

*I wonder.* Does the investigation connect to a student's personal interests? Or does the outcome suggest a question or pique a student's curiosity? Providing time for students to write "I wonder" statements or questions supports the idea that the pursuit of scientific knowledge does not end with the day's investigation. The notebook is an excellent place to capture students' musings and for students to record thoughts that might otherwise be lost.

*Wrap-up/warm-up.* At the end of each investigation part or at the beginning of the next part, students will engage in a wrap-up or warm-up. This is another opportunity for students to revisit the content of the investigation. As partners discuss their responses to the focus question, they may choose to add information to their responses. This should be encouraged.

▶ **NOTE**
More information about partner and small-group discussion protocols that can be used during the Wrap-Up/Warm-Up session appears in the Science-Centered Language Development chapter.

---

## 21. Share notebook entries

Conclude Part 1 or start Part 2 by having students share notebook entries. Ask students to open their science notebooks to the most recent entry. Read the focus question together.

➤ *What is soil?*

Ask students to pair up with a partner to

- share their answers to the focus question;
- discuss the different kinds of materials that were found in the four different soils;
- share their ideas about which soil came from which location—desert, river delta, mountain, forest.

---

## Next-Step Strategies

In each investigation, the *Investigations Guide* indicates an assessment opportunity and what to look for when examining students' work. The purpose of looking at students' work at intermediate junctures is for embedded (formative) assessment, *not* for grading. Look for patterns in students' understanding by collecting and sorting the notebooks. If the patterns indicate that students need additional help with communication or with content, you might want to select a corrective next-step strategy before going on to the next part. This process of looking critically at students' work is described in more detail in the Assessment chapter.

A next-step strategy is a brief instructional activity that takes place before the start of the next investigation part. Select a strategy based on students' needs to communicate their thinking more efficiently or accurately, use scientific vocabulary effectively, or think about the concept in a different way.

During each next step, students are engaged in reflection and self-assessment. Scientists constantly refine and clarify their ideas about how the natural world works. They read scientific articles, consult with other scientists, and attend conferences. They incorporate new information into their thinking about the subject they are investigating. This reflective process can result in deeper understanding or even a complete revision in thinking.

Likewise, when students receive additional instruction or information after writing a conclusion, response sheet, or answer to a focus question, give them time to reflect on how their ideas might have changed. In the self-assessment that follows, they review their original written work, judge its accuracy and completeness, and write a revised explanation.

What follows here is a collection of next-step strategies that teachers have used successfully with groups of students to address areas of need. The strategies are flexible enough to use in different groupings and can be modified to meet your students' needs.

By engaging in any of these next-step strategies, students have to think actively about every aspect of their understanding of the concept and organize their thoughts into coherent, logical narratives. The learning that takes place during this reflection process is powerful. The relationships between the several elements of the concept become unified and clarified.

---

**TEACHING NOTE**

*Students' learning can be assessed only at the level in which it was done. If students worked in groups to answer the focus question, it is difficult to assess individual understanding. Similarly, providing a frame to guide a student response provides evidence on what students can do at a supported level, not at an independent level.*

### Next-Step Strategies

- *Teacher feedback*
- *Line of learning*
- *Review and critique*
- *Key points*
- *Revision with color*
- *Class debate*
- *Mini-lessons*

**Teacher feedback.** Students' writing often exposes weaknesses in students' understanding—or so it appears. It is important to check whether the flaw results from poor understanding of the science or from imprecise communication. You can use the notebook to provide feedback, asking for clarification or additional information. Attach a self-stick note, which can be removed after the student has taken appropriate action. The most effective forms of feedback relate to the content of the work. Nonspecific feedback (such as stars, smiley faces, and "good job!") and ambiguous critiques (such as "try again," "put more thought into this," and "not enough") are less effective. Here are a couple of examples.

- *You claim that water condensed on the glass of ice water. Where did the water come from?*

- *Tell me why you think Lightbulb A will light. Hint: Check the contact points on the two illustrations.*

Feedback that guides students to think about the content of their work and gives suggestions for how to improve are productive instructional strategies. Here are some examples of useful generic feedback.

- *Use the science vocabulary in your answer.*

- *Include an example to support your ideas.*

- *Include more detail about _____.*

- *Check your data to make sure this is right.*

- *Include units with your measurements.*

When students return to their notebooks and respond to the feedback, you will have additional information to help you discriminate between knowledge and communication difficulties.

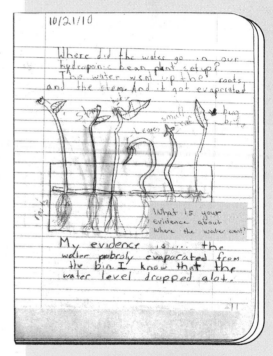

**Students can respond to feedback using a different-color pencil.**

**Line of learning.** One technique many teachers find useful in the reflective process is the line of learning. After students enter their initial explanation, followed by discussion, assessment, reading, and/or teacher feedback, they draw and date a line under their original work. Then they make a new entry under the line of learning, adding to or revising their original thinking. If the concept is elusive or complex, a second line of learning, followed by more processing and revising, may be appropriate.

The line of learning is a reminder to students that learning is an ongoing process, and that the products of the process are imperfect. The line of learning marks places in that process where a student made a stride toward full understanding. The psychological security provided by the line of learning reminds students that they can always draw another line of learning and revise their thinking once again. The ability to look back in the science notebook and see concrete evidence of learning gives students confidence and helps them become critical observers of their own learning.

**Review and critique anonymous student work.** Presenting work from other students can be a valuable learning tool for refining and improving responses for content and literacy. Depending upon the culture of the class, you might present actual or simulated student work from a focus question or response sheet, selected to represent a common misconception, error, or exemplary work. Present it to the whole class, and then have students work in groups to discuss the merits of the response and make recommendations for improvement. In this process, students discuss what information is needed in a quality response. After critiquing other students' responses, students look at their own responses, draw a line of learning, and refine their own thinking.

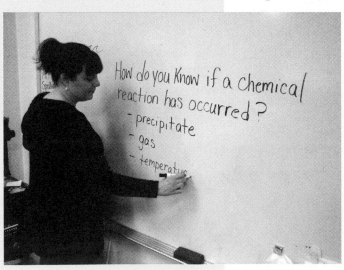

**Key points.** As students advance in years, so does the level of questions they are asked to answer. Some questions require students to address several points for a complete response. If students discuss only one point well, it may be that they are unaware of the other points or need assistance in how to address multiple points in an answer. Using key points, you pose the question to the group and, through discussion, elicit the key ideas or points that would be included in a complete answer to the question. On the board, record only words or brief phrases. Once the class has agreed on the key points, students review their responses and add information to answer the question completely. The list of words and phrases provides the support for revision, but not the language.

**Revision with color.** Suppose a student response of several sentences has some accurate information, some information that needs refinement, and a critical gap. Instead of crossing out the original response completely and starting anew, students can use color to revise their responses. After discussing the question with a partner or in a group, or even during key points, students grab three different-color pens or pencils and refine their responses by using the three Cs (confirm, correct, and complete). To confirm information as accurate, they underline it in green. If they need to correct a misconception, they do so in red. If they need to complete their responses by adding more information, they do it in blue. In doing so, students learn that their initial responses are works in progress, and, like scientists, students may revisit and modify their responses based on new information.

**Class debate.** If students have differing points of view on a response, you could have students engage in a healthy class debate. After establishing rules to foster a respectful and helpful class environment, volunteers read their answers, and classmates can agree or disagree. Students can present supporting evidence or counterarguments. An important part of the debate is that students may change their minds once they have additional evidence. Students return to their notebooks and can modify their responses by using any other next step strategy.

**Mini-lessons.** Sometimes the data from sorting notebook entries reveal that students need some additional information or specific guidance on a skill. A mini-lesson is a brief interaction with a group of students, targeting the area of need. You might have a group of students observe a stream table more closely in order to observe landforms or give a writing prompt to a small group of students and ask them to explain their thinking more clearly.

# WRITING OUTDOORS

Every time you go outdoors with students, you will have a slightly different experience. Naturally, the activity or task will be different, but other variables may change as well. The temperature, cloud cover, precipitation, moisture on the ground; other activities unexpectedly happening outside; students' comfort levels related to learning outdoors; and time of the school year are all aspects that could affect the activity and will certainly determine how you incorporate the use of notebooks. Most students are completely capable of staying on task with the outdoor experience when they are doing science, but writing outdoors can be a bit trickier. The following techniques are tried-and-true ways to help students learn how to write outdoors and to give them all the supplies they need to support their writing.

## Create "Desks"

Students need a firm writing surface. Students who write in composition notebooks with firm covers can simply fold them open to the pages they are writing on, rest them in the crook of their nonwriting arms, and hold them steady with their nonwriting hands—they can stand, sit, kneel, or lean against a wall to write. At the beginning of the school year, take a minute to model how to do this.

Many students feel most comfortable sitting down to write. Curbs, steps, wooden stumps or logs, rocks, and grass are places to sit while

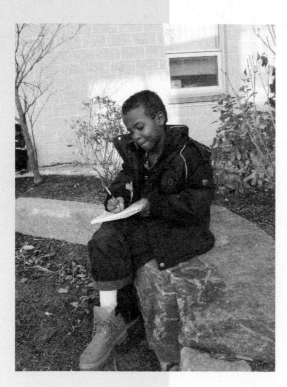

writing. Select a writing location that suits your students' comfort levels. Some students will not be comfortable sitting on the grass or ground. They will need to sit on something such as a curb, stair, boulder, or stump at the start of the year, but will eventually feel more comfortable with all aspects of the outdoor setting as the year moves along.

If students are using individual notebook sheets or notebooks with flimsy covers, you will likely want to buy or make clipboards. If you do not have clipboards, use a box cutter to cut cardboard to the proper size. Clamp a binder clip at the top to make a lightweight yet sturdy clipboard. If it gets ruined, no tears will be shed. If you're in the market for new clipboards, get the kind that are stackable and do not have a bulky clip. Ideally, all the clipboards will fit in one bag for portability and easy distribution.

If using a notebook sheet, simply put the sheet on the clipboard before going outdoors, and have students glue the sheet into their notebooks when you are back in the classroom. If you are using three-ring binders, do not bring them outdoors. Take the paper out, put it on a clipboard, and return it to the binder when back in the classroom. Don't take the risk of the rings opening and everything blowing away.

Elastic bands around the bottom of the clipboard, or around a stiffer composition notebook, will help keep the paper from flapping around and becoming too weathered.

## Bring Writing Tools Outdoors

You can bring chart paper outdoors. Roll up a blank piece of chart paper, grab some blue painter's tape, and stick the piece of paper to the school wall. You'll need to tape all four corners. Or set up a chart inside, and clip it to a chain-link fence with binder clips or clothespins when you go outdoors.

Take along extra pencils, as pencil points will break. Some teachers find it helpful to tie pencils onto clipboards. Pencils should be tucked between the clip and the notebook sheet so that students don't poke themselves or, more likely, accidentally break the pencil points.

## Decide When to Write Outdoors

In general, notebook entries will be more detailed and more insightful if students can stay outdoors where the scientific exploration occurred. Sometimes, you will want to complete notebook entries indoors. If you are teaching the module early in the year when students are building up the routines for using the schoolyard, or if the weather is not ideal (a little chilly, raining, too hot, too windy), then you may want students to make notebook entries after returning to the classroom. If students are totally focused and in the moment, they can stay outdoors while they answer the focus question. If other students are outdoors playing, you may need to bring the class indoors to complete the written work. Only you will be able to determine what is best at the time.

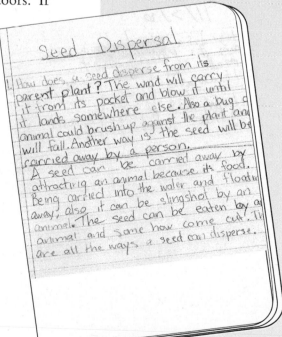

# CLOSING THOUGHTS

Engaging students in active science with notebooks provides a rich experience. Doing this successfully requires thoughtful interactions among students, materials, and natural phenomena. Initially, adding notebooks to your science teaching will require you to focus students' attention on how to set up the notebook, what types of entries students should make, and when students should be using their notebooks. You will establish conventions about where to record the date and title, where to keep notebooks, how to glue notebook sheets into notebooks, and when to record observations and thinking.

Once you are past these perfunctory issues, you can shift your focus to the amount of scaffolding to provide to students or to encouraging students to create their own notebook entries. During this time, you and your students are developing skills to improve the quality of notebook entries. These skills may include asking better questions to focus students' attention on a specific part of an organism or using color to enhance a drawing. Students begin to make entries with less prompting. They give more thought to supporting their responses to the focus question. When asked to make a derivative product, students thumb through their notebooks to find the needed information. The notebook becomes a tool for students to help recall their learning.

As students begin to document their thinking about focus questions and other queries, you may begin to wonder, "Should I be doing something with their notebooks?" This is when your focus shifts from the notebook as just something students use during science learning to the notebook as an assessment tool. Once everyone is comfortable recording the focus question and collecting data, you can take the next step of collecting notebooks and reading students' responses as a measure of not just how individual students are learning, but what the pervasive needs of students are. You choose next-step strategies that address students' needs before proceeding to the next investigation. The notebooks act as an assessment tool that lets you modify your science instruction.

This process will take time, discussions with colleagues, revisiting different sections of this chapter, and critical scrutiny of students' work before both you and your students are using notebooks to their full potential.

> ▶ **NOTE**
> For more on derivative products, see the Science-Centered Language Development chapter.

# Science-Centered Language Development

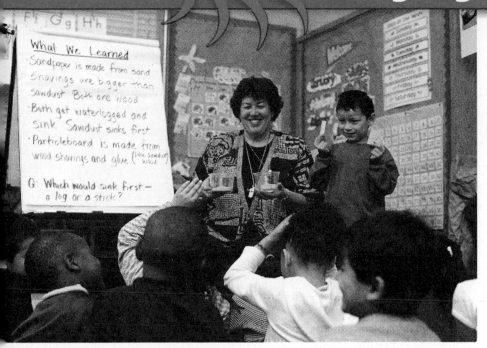

**Contents**

Introduction ............................1

The Role of Language in
Scientific and Engineering
Practices ...............................3

Speaking and
Listening Domain ...................**6**

Writing Domain...................**12**

Reading Domain ..................**21**

Science-Vocabulary
Development........................**30**

English-Language
Development ......................**36**

References ..........................**42**

*Teams of science inquirers talk about and write about their questions, their tentative explanations, their relationships between evidence and explanations, and their reasons and judgments about public presentations and scientific arguments in behalf of their work. It is in the context of this kind of scientific activity that students' literacy of the spoken and written word develops along with the literacy of the phenomenon.*

Hubert M. Dyasi, "Visions of Inquiry: Science"

# INTRODUCTION

In this chapter, we explore the intersection of science and language and the implications for effective science teaching and language development. We identify best practices in language arts instruction that support science learning and examine how learning science content and practices supports language development. The active investigations, science notebooks, *FOSS Science Resources* readings, and formative assessment activities in FOSS provide rich contexts in which students develop and exercise thinking processes and communication skills. Together, these elements comprise effective instruction in both science and language arts—students experience the natural world around them in real and authentic ways and use language to inquire, process information, and communicate their thinking about scientific phenomena. We refer to the development of language within the context of science as science-centered language development.

# Science-Centered Language Development

Language plays two crucial roles in science learning: (1) it facilitates the communication of conceptual and procedural knowledge, questions, and propositions (external, public), and (2) it mediates thinking, a process necessary for understanding (internal, private). These are also the ways scientists use language: to communicate with one another about their inquiries, procedures, and understandings; to transform their observations into ideas; and to create meaning and new ideas from their work and the work of others. For students, language development is intimately involved in their learning about the natural world. Science provides a real and engaging context for developing literacy, and language arts skills and strategies support conceptual development and scientific practices. For example, the skills and strategies used for reading comprehension, writing expository text, and oral discourse are applied when students are recording their observations, making sense of science content, and communicating their ideas. Students' use of language improves when they discuss, write, and read about the concepts explored in each investigation.

We begin our exploration of science and language by focusing on language functions and how specific language functions are used in science to facilitate information acquisition and processing (thinking). Then we address issues related to the specific language domains—speaking and listening, writing, and reading. Each section addresses

- how skills in that domain are developed and exercised in FOSS science investigations;

- literacy strategies that are integrated purposefully into the FOSS investigations;

- suggestions for additional literacy strategies that both enhance student learning in science and develop or exercise English-language literacy skills.

Following the domain discussions is a section on science-vocabulary development, with scaffolding strategies for supporting all learners. The last section covers language-development strategies specifically for English learners.

▶ **NOTE**
The term *English learners* refers to students who are learning to understand English. This includes students who speak English as a second language and native English speakers who need additional support to use language effectively.

# THE ROLE OF LANGUAGE IN SCIENTIFIC AND ENGINEERING PRACTICES

Language functions are the purpose for which speech or writing is used and involve both vocabulary and grammatical structure. Understanding and using language functions appropriately is important in effective communication. Students use numerous language functions in all disciplines to mediate communication and facilitate thinking (e.g., they plan, compare, discuss, apply, design, draw, and provide evidence).

In science, language functions facilitate the enactment of scientific and engineering practices. For example, when students are *collecting data*, they are using language functions to identify, label, enumerate, compare, estimate, and measure. When students are *constructing explanations*, they are using language functions to analyze, communicate, discuss, evaluate, and justify.

*A Framework for K–12 Science Education: Practices, Crosscutting Concepts, and Core Ideas* (National Research Council 2012) provides a "set of essential practices as essential for any education in the sciences and engineering." Each of these scientific and engineering practices requires the use of multiple language functions. Often, these language functions are part of an internal dialogue weighing the merits of various explanations—what we call thinking. The more language functions with which we are facile, the more effective and creative our thinking can be.

The scientific and engineering practices are listed below, along with a sample of the language functions that are exercised when effectively engaged in that practice. (Practices are bold; language functions are italic.)

## Asking questions and defining problems

- *Ask* questions about objects, organisms, systems, and events in the natural world (science) or define a problem (engineering).

## Planning and carrying out investigations; analyzing, and interpreting data

- *Plan* and conduct investigations.
- *Observe* and *record* data.
- *Measure* to extend the senses to acquire data.
- *Organize* observations (data), using numbers, words, images, and graphics.

**Examples of Language Functions**

- Analyze
- Apply
- Ask questions
- Clarify
- Classify
- Communicate
- Compare
- Conclude
- Construct
- Critique
- Describe
- Design
- Develop
- Discuss
- Distinguish
- Draw
- Enumerate
- Estimate
- Evaluate
- Experiment
- Explain
- Formulate
- Generalize
- Group
- Identify
- Infer
- Interpret
- Justify
- Label
- List
- Make a claim
- Measure
- Model
- Observe
- Organize
- Plan
- Predict
- Provide evidence
- Reason
- Record
- Represent
- Revise
- Sequence
- Solve
- Sort
- Strategize
- Summarize

## Constructing explanations (science) and designing solutions (engineering); engaging in argument from evidence

- *Predict* future events, based on evidence and reasonings.
- Use data and logic to *construct* and *communicate* reasonable explanations.
- *Develop*, *discuss*, *evaluate*, and *justify* the merits of explanations.
- Use science-related technologies to *apply* scientific knowledge and *solve* problems.

## Developing and using models

- Create models to *explain* natural phenomena and *predict* future events or outcomes.
- *Describe* how models represent the natural world.
- *Compare* models to actual phenomena and *identify* limitations of the models.

## Obtaining, evaluating, and communicating

- *Identify* key ideas in text, identify supporting evidence, and know how to use various text features.
- *Distinguish* opinion from evidence.
- Use writing as a tool for *clarifying* ideas and *communicating*.

Research supports the claim that when students are intentionally using language functions in thinking about and communicating in science, they improve not only science content knowledge, but also language-arts and mathematics skills (Ostlund 1998; Lieberman and Hoody 1998). Language functions play a central role in science as a key cognitive tool for developing higher-order thinking and problem-solving abilities that, in turn, support academic literacy in all subject areas.

Here is an example of how an experienced teacher can provide an opportunity for students to exercise language functions in FOSS. In the **Soils, Rocks, and Landforms Module**, this is one piece of content we expect students to have acquired by the end of the module.

- The greater the flow of water across Earth's surface, the greater the rate of erosion and deposition.

The scientific practices the teacher wants the class to focus on are *interpreting data* and *constructing explanations*.

The language functions students will exercise while engaging in these scientific practices are *comparing, explaining,* and *providing evidence*. The teacher understands that these language functions are appropriate to the purpose of the science investigation and support the Common Core Standards for writing (students will write explanatory texts to examine a topic and convey ideas and information clearly).

- Students will *compare* observational data (from a stream-table investigation) to *explain* the relationship between the amount of water that runs over a surface and the amount of erosion and deposition that occur.

The teacher can support the use of language functions by providing structures such as sentence frames.

- As \_\_\_\_\_, then \_\_\_\_\_.

  *As more water flows over a surface, then more erosion and deposition occur.*

- When I changed \_\_\_\_\_, then \_\_\_\_\_ happened.

  *When I changed the amount of water flowing down the stream table, then more erosion and deposition happened.*

- The more/less \_\_\_\_\_, the \_\_\_\_\_.

  *The more water that flows across the earth materials, the more erosion and deposition that occur.*

▶ **NOTE**
For more examples of how FOSS teachers address language-arts standards while conducting science investigations with students, go to FOSSweb (www.FOSSweb.com).

## SPEAKING AND LISTENING DOMAIN

The FOSS investigations are designed to engage students in productive oral discourse. Talking requires students to process and organize what they are learning. Listening to and evaluating peers' ideas calls on students to apply their knowledge and to sharpen their reasoning skills. Guiding students in small-group and whole-class discussions is critical to the development of conceptual understanding of the science content and the ability to think and reason scientifically.

FOSS investigations start with a discussion—either a review to activate prior knowledge or presentation of a focus question or a challenge to motivate and engage active thinking. During active investigation, students talk with one another in small groups, share their observations and discoveries, point out connections, ask questions, and start to build explanations. The discussion icon in the sidebar of the *Investigations Guide* indicates when small-group discussions should take place.

During the activity, the *Investigations Guide* indicates where it is appropriate to pause for whole-class discussions to guide conceptual understanding. The *Investigations Guide* provides you with discussion questions to help stimulate student thinking and support sense making. At times, it may be beneficial to use sentence frames or standard prompts to scaffold the use of effective language functions and structures.

At the end of the investigation, there is another opportunity to develop language through speaking and listening in the Wrap-Up/Warm-Up. During this time, students are often asked to discuss their responses to the focus question. All students need to be engaged in this opportunity to use different language functions.

On the following pages are some suggestions for providing structure to those discussions and for scaffolding productive discourse when needed. Teaching techniques used to generate discussion in language arts and other content areas can also be used effectively during science. Using the protocols that follow will ensure inclusion of all students in discussions.

# Partner and Small-Group Discussion Protocols

Whenever possible, give students time to talk with a partner or in a small group before conducting a whole-class discussion. This provides all students with a chance to formulate their thinking, express their ideas, practice using the appropriate science vocabulary, and receive input from peers. Listening to others communicate different ways of thinking about the same information from a variety of perspectives helps students negotiate the difficult path of sense making for themselves.

**Dyads.** Students pair up and take turns either answering a question or expressing an idea. Each student has 1 minute to talk while the other student listens. While student A is talking, student B practices attentive listening. Student B makes eye contact with student A, but cannot respond verbally. After 1 minute, the roles reverse.

Here's an example from the **Mixtures and Solutions Module**. Just before students answer the focus question in their notebooks, you ask students to pair up and take turns sharing their answer to the question "How do you know when a chemical reaction has occurred?" The language objective is for students to be able to explain what happens during a chemical reaction and provide evidence (orally and in writing) that it has occurred. These sentence frames can be written on the board to scaffold student thinking and conversation.

- I claim that _____ produces a chemical reaction, because _____.

- Evidence for my claim includes _____.

**Partner parade.** Students form two lines facing each other. Present a question, an idea, an object, or an image as a prompt for students to discuss. Give students 1 minute to greet the person in front of them and discuss the prompt. After 1 minute, call time. Have the first student in one of the lines move to the end of the line, and have the rest of the students in that line shift one step sideways so that everyone has a new partner. (Students in the other line do not move.) Give students a new prompt to discuss for 1 minute with their new partners.

For example, students are just beginning the investigation on landforms, and you want to assess prior knowledge. Give each student a picture of a landform, and have students line up facing each other. For the first round, ask, "What do you know about the image on your card?" For the second round, ask, "How do you think the landform in your picture formed?" For the third round, ask, "What personal experience have you had with this landform?" The language objective is for students to describe their observations, infer how the landform formed, and reflect upon and relate any experiences they may have had with a similar landform. The following sentence frames can be used to scaffold student discussion.

- I notice _____.

- I think the landform was formed by _____.

- The landform in my picture reminds me of the time when _____.

**Put in your two cents.** For small-group discussions, give each student two pennies or similar objects to use as talking tokens. Each student takes a turn putting a penny in the center of the table and sharing his or her idea. Once all have shared, each student takes a turn putting in the other penny and responding to what others in the group have said. For example,

- I agree (or don't agree) with _____ because _____.

Here's an example from the **Sun, Moon, and Planets Module**. Students have been monitoring their shadows throughout the day and are still struggling to make coherent connections between Earth's rotation and the effect that it has on shadows. The language objective is for students to describe their observations, explain why the direction and length of their shadows change during the day, and provide evidence based on their own observations. You give each student two pennies, and in groups of four, they take turns putting in their two cents. For the first round, each student answers the question "Why do shadows change during the day?" They use the frame

- I think shadows change during the day because _____.

- My evidence is _____.

On the second round, each student states whether he or she agrees or disagrees with someone else in the group and why, using the sentence frame.

## Whole-Class Discussion Supports

The whole-class discussion is a critical part of sense making. After students have had the active learning experience and have talked with their peers in partners and/or small groups, sharing their observations with the whole class sets the stage for developing conventional explanatory models. Discrepant events, differing results, and other surprises are discussed, analyzed, and resolved. It is important that students realize that science is a process of finding out about the world around them. This is done through asking questions, testing ideas, forming explanations, and subjecting those explanations to logical

**Whole-Class Discussion Supports**
- *Sentence frames*
- *Guiding questions*

scrutiny. Leading students through productive discussion helps them connect their observations and the abstract symbols (words) that represent and explain those observations. Whole-class discussion also provides an opportunity for you to interject an accurate and precise verbal summary statement as a model of the kind of thinking you are seeking. Facilitating effective whole-class discussions takes skill, practice, a shared set of norms, and patience. In the long run, students will have a better grasp of the content and improve their abilities to think independently and communicate effectively.

Norms should be established so that students know what is expected during science discussions.

- Science content and practices are the focus.

- Everyone participates.

- Ideas and experiences are shared, accepted, and valued. Everyone is respectful of one another.

- Claims are supported by evidence.

- Challenges (debate and argument) are part of the quest for complete understanding.

The same discussion techniques used during literacy circles and other whole-class discussions can be used during science instruction (e.g., attentive listening, staying focused on the speaker, asking questions, responding appropriately). In addition, in order for students to develop and practice their reasoning skills, they need to know the language forms and structures and the behaviors used in evidence-based debate and argument (e.g., using data to support claims, disagreeing respectfully, asking probing questions; Winokur and Worth 2006).

Explicitly model and conduct mini-lessons (5 to 10 minutes of focused instruction) on the language structures appropriate for active discussions, and provide time for students to practice them, using the science content.

**Sentence frames.** The following samples can be posted as a scaffold as students learn and practice their reasoning and oral participation skills.

- I think _____, because _____.

- I predict _____, because _____.

- I claim _____; my evidence is _____.

- I agree with _____ that _____.

- My idea is similar/related to _____'s idea.

> **TEACHING NOTE**
>
> Let students know that scientists change their minds based on new evidence. It is expected that students will revise their thinking based on evidence presented in discussions.

> **TEACHING NOTE**
>
> Encourage science talk. Allow time for students to engage in discussions that build on other students' observations and reasoning. After an investigation, use a teacher- or student-generated question, and either just listen or facilitate the interaction with questions to encourage expression of ideas among students.

- I learned/discovered/heard that _____.

- <Name> explained _____ to me.

- <Name> shared _____ with me.

- We decided/agreed that _____.

- Our group sees it differently, because _____.

- We have different observations/results. Some of us found that _____. One group member thinks that _____.

- We had a different approach/idea/solution/answer _____.

**Guiding questions.** The *Investigations Guide* provides questions to help concentrate student thinking on the concepts introduced in the investigation. Guiding questions should be used during the whole-class discussion to facilitate sense making. Here are some other open-ended questions that help guide student thinking and promote discussion.

- What did you notice when _____?

- What do you think will happen if _____?

- How might you explain _____? What is your evidence?

- What connections can you make between _____ and _____?

# Whole-Class Discussion Protocols

The following examples of tried-and-true participation protocols can be used to enhance whole-class discussions during science and all other curriculum areas. The purpose of these protocols is to increase meaningful participation by giving all students access to the discussion, allowing students time to think (process), and providing a context for motivation and engagement.

**Think-pair-share.** When asking for a response to a question posed to the class, allow time for students to think silently for a minute. Then, have students pair up with a partner to exchange thoughts before you call on a student to share his or her ideas with the whole class.

**Pick a stick.** Write each student's name on a craft stick, and keep the sticks handy at the front of the room. When asking for responses, randomly pick a stick, and call on that student to start the discussion. Continue to select sticks as you continue the discussion. Your name can also be on a stick in the cup. You can put the selected sticks in a different location or back into the same cup to be selected again.

**Whip around.** Each student takes a quick turn sharing a thought or reaction. Questions are phrased to elicit quick responses that can be expressed in one to five words (e.g., "Give an example of a stored-energy source." "What does the word *heat* make you think of?").

**Group posters.** Have small groups design and graphically record their investigation data and conclusions on a quickly generated poster to share with the whole class.

**Two-cup pick-a-stick container**

**One-cup pick-a-stick container divided with tape**

# WRITING DOMAIN

Information processing is enhanced when students engage in informal writing. When allowed to write expressively without fear of being scorned for incorrect spelling or grammar, students are more apt to organize and express their thoughts in different ways that support sense making. Writing promotes the use of scientific practices, thereby developing a deeper engagement with the science content. It also provides guidance for more formal derivative science writing (Keys 1999).

## Science Notebooks

The science notebook is an effective tool for enhancing learning in science and exercising various forms of writing. Science notebooks provide opportunities both for expressive writing (students craft explanatory narratives that make sense of their science experiences) and for practicing informal technical writing (students use organizational structures and writing conventions). Starting as emergent writers in kindergarten, students learn to communicate their thinking in an organized fashion while engaging in the cognitive processes required to develop concepts and build explanations. Having this developmental record of learning also provides an authentic means for assessing students' progress in both scientific thinking and communication skills.

One way to help students develop the writing skills necessary for productive notebook entries is to focus on the corresponding language functions. The language forms and structures used to perform these language functions in science are used in all curricular areas and, therefore, can be suitably taught in conjunction with existing language-arts instruction. This can be done through mini-lessons on the writing skills that support the various types of notebook entries.

Table 1, at the end of the Writing Domain section, provides examples of how language functions are used to help students develop both their general writing skills and their thinking abilities within the format of the science-notebook entry. The writing objectives for a mini-lesson, along with the particular language forms and structures (vocabulary, syntax, linking words, organization of ideas, and so on), are identified in the table along with the suggested sentence frames for scaffolding.

For example, if students are observing a particular insect's behavior, the language objective might be "Students describe in detail the behavior of the insect." A prior mini-lesson on using sensory details to describe observations would provide students with the language forms and structures appropriate for recording their data in their notebooks during the observations. As a scaffold, students could also be provided with sentence frames to help them write detailed narratives.

> **NOTE**
> For more information about supporting science-notebook development, see the Science Notebooks chapter.

> **NOTE**
> Language forms and structures refer to the internal grammatical structure of words and how those words go together to make sentences.

| Language function | Language objectives for writing in notebooks | Language forms, structures, and scaffolds for writing |
|---|---|---|
| Notebook component—data acquisition | | |
| Describe | Write narratives: use details, sensory observations, and connections to prior knowledge. | I observed ___. When I touch the ___, I feel ___. The ___ has ___. I noticed ___. It feels ___. It smells ___. It sounds ___. It reminds me of ___, because ___. |

This sample from Table 1 shows how language functions can be developed and applied when writing in science notebooks.

## Developing Derivative Language-Arts Products

Science notebooks provide students with a source of information (content) from which they can draw to create more formal science-centered writing pieces. Derivative products are written pieces that are generated with specific language-arts goals in mind, such as audience and language function. Writing-to-learn methods can enhance science concepts when students engage in different types of writing for different purposes (Hand and Prain 2002). We know that students are more engaged and motivated to write when they have a clear and authentic context for writing.

The language extensions in the Interdisciplinary Extensions section at the end of each investigation suggest writing activities that can be used to help students learn the science content for that particular investigation. The writing activities incorporate language-arts skills appropriate for the grade level. Questions and ideas for future writing activities that surface during the investigations can be recorded in a class list, in science notebooks, or in students' writing folders. Here are general suggestions for using science content to create products in each of several writing genres.

***Descriptive writing.*** Students use descriptive writing to portray an organism, an environment, an object, or a phenomenon in such a way that the reader can almost recapture the writer's experience. This is done through the use of sensory language; rich, vivid, and lively detail; and figurative language, such as simile, hyperbole, metaphor, and symbolism. You can remind students to show, rather than tell, through the use of active verbs and precise modifiers.

To help them learn the science content, students can use the information in their science notebooks to elaborate on their observations by using descriptive vocabulary, analogies, and drawing.

▶ **NOTE**
The complete table appears at the end of this Writing Domain section.

**Derivative Language-Arts Products**
- *Descriptive writing*
- *Persuasive writing*
- *Narrative writing*
- *Expository writing*
- *Recursive cycle*

For example, kindergartners can draw pictures of isopods, fish, worms, things made of wood, different types of trees and leaves, and so on. Through drawings and words, students describe objects' properties and compare how those objects are the same and different.

Primary students can make property cards by writing on an index card as many properties as they can that describe an object or organism. Then, students take turns reading the properties to another student to see if the partner can identify the corresponding object.

Students of all ages can expand on their notebook entries in poetry that expresses their interpretation of organisms, objects, and phenomena such as crayfish, seeds, electricity, rivers, weather, and minerals. The use of similes is a good way to engage students in making comparisons. You can use a simple frame, such as: _____ is like _____. (For example, in the **Air and Weather Module**, students might write "The weather today is like a dragon.")

Students can share their similes. Other students can explain why they think the simile works.

***Persuasive writing.*** The objective of persuasive writing is to convince the reader that a stated opinion or interpretation of data is worthwhile and meaningful. Students learn to support their claims with evidence and to use persuasive techniques, such as logical arguments and calls to action. By using claims and evidence to formulate conclusions in their science notebooks, students develop and apply their thinking and reasoning skills to form the basis of persuasive writing in a variety of formats, such as essays, letters, editorials, advertisements, award nominations, informational pamphlets, and petitions. Animal habitats, energy use, weather patterns, landforms, and water sources are just a few science topics that can generate questions and issues for persuasive writing.

Here is a sample of persuasive writing frames (modified from Gibbons 2002).

Title: _____
The topic of this discussion is _____.
My opinion (position, conclusion) is _____.
There are <number> reasons why I believe this to be true.
First, _____.
Second, _____.
Finally, _____.
On the other hand, some people think _____.
I have also heard people say _____.
However, my claim is that _____ because _____.

**Narrative writing.** Science provides a broad landscape of engaging material for stimulating the imagination for the writing of stories, songs, biographies, autobiographies, poems, and plays. Students can use organisms or objects as characters; describe habitats and environments as settings; and write scripts portraying various systems, such as weather patterns, flow of electricity, and water, rock, or life cycles.

**Expository writing.** Students use science content to inform, explain, clarify, define, or instruct through writing letters, definitions, procedures, newspaper and magazine articles, posters, pamphlets, and research reports. Expository writing is characterized by a focus on main topics with supporting facts, details, explanations, and examples and is organized in a clear, coherent, and sequential manner. Expository writing has a clear focus on communicating accurate, complete, and detailed representations of science content and/or observation of natural history or events.

During writing instruction, students can use the information in their science notebooks and in related readings (and other sources, such as video content) to write a more formal and conclusive answer to the focus question. Strategies such as the writing process (plan, draft, edit, revise, and share) and writing frames (modeling and guiding the use of topic sentences, transition and sequencing words, examples, explanations, and conclusions) can be used with the science content to develop proficiency in critical writing skills.

▶ **NOTE**
Human characteristics should not be given to organisms (anthropomorphism) in science investigations, only in literacy extensions.

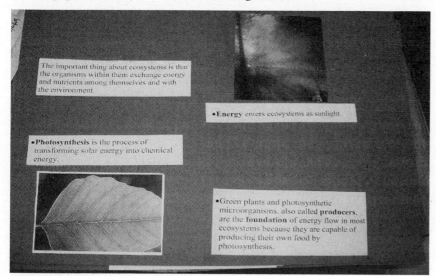

A page from a big book about ecosystems, a derivative product created during language arts

Here are three samples of writing frames for science reports for primary and upper-elementary students (modified from Wellington and Osborne 2001).

**Primary** (life cycle appropriate for the **Insects and Plants Module**)

Title: _____

(Define) A _____ is _____.

(Classify) A _____ is a kind of _____.

(Describe) A _____ is _____.

(Habitat) _____ lives _____.

(Life cycle) _____ started life as _____.

(Change over time) Then it _____.

(Food) A _____ eats _____.

(Evidence) I know that _____ because _____.

**Upper elementary** (living structures appropriate for the **Structures of Life Module**)

Title: _____

(Identify) The part of the body I am describing is the _____.

(Describe) It consists of _____.

(Explain) The function of these parts is _____.

(Example) This drawing shows _____.

**Upper elementary** (explanation)

Title: _____

I want to explain why (how) _____.

An important reason for why (how) this happens is that _____.

Another reason is that _____.

I know this because _____.

**Recursive cycle.** An effective method for extending students' science learning through writing is the recursive cycle of research (Bereiter 2002). This strategy emphasizes writing as a process for learning, similar to the way students learn during the active science investigations.

1. Decide on a problem or question to write about.

2. Formulate an idea or a conjecture about the problem or question.

3. Identify a remedy or an answer, and develop a coherent discussion.

4. Gather information (from science notebooks, *FOSS Science Resources*, books, FOSSweb interactives, Internet, interviews, videos, experiments, etc.).

5. Reevaluate the problem or question based on what has been learned.

6. Revise the idea or conjecture.

7. Make presentations (reports, posters, electronic presentations, etc.).

8. Identify new needs, and make new plans.

This process can continue for as long as new ideas and questions occur, or students can present a final product in any of the suggested formats.

**Table 1.** Examples of how language functions are exercised in science notebook writing and examples of sentence frames and language structures students may use

| Language function | Language objectives for writing in notebooks | Language forms, structures, and scaffolds for writing |
|---|---|---|
| **Organizational setup** | | |
| Organize | Set up and organize notebook: table of contents, glossary or index, page numbers, date, turning to the next blank page. | Use structures such as rows, columns, blocks, numbers, location, alphabetizing. |
| **Notebook component—planning the investigation** | | |
| Design Strategize | Write a narrative plan: communicate ideas on an approach to answer the focus question or challenge posed in the investigation. | Use adverbs such as *first*, *second*, *next*, *then*, *finally*. Brainstorming ideas: First I will _____, and then I will _____. I will need to _____ to _____. |
| List Record | Write ordered lists: materials, variables, vocabulary words, bullets | We need _____, _____, _____, and _____ to _____ . |
| Sequence | Record step-by-step procedures: explain exactly what to do, number steps in a sequence, specifics for measurement, how controls will be established, how variables will be measured (tools/units). | 1. _____. 2. _____. 3. _____. I will change _____. I will not change _____. I will measure _____. I will not measure _____. |
| **Notebook component—data acquisition** | | |
| Describe | Write narratives: use details, sensory observations, connections to prior knowledge. | I observed _____. When I touch the _____, I feel _____. The _____ has _____. I noticed _____. It feels _____. It smells _____. It sounds _____. It reminds me of _____, because _____. |

**Table 1 (*continued*)**

| Language function | Language objectives for writing in notebooks | Language forms, structures, and scaffolds for writing |
|---|---|---|
| **Notebook component—data acquisition** (continued) | | |
| Draw<br>Label<br>Identify | Make technical drawings: draw large, accurate, and detailed representations; identify parts of a system. | Label drawing, using science vocabulary. Recognize shapes, form, location, color, size, and scale.<br><br>My drawing shows ____. |
| Organize<br>Compare<br>Classify | Make charts and tables: use a T-table or chart for recording and displaying data. | Set up rows, columns, headings.<br><br>My T-table compares ____. |
| Sequence<br>Compare | Record changes: use language structures to communicate change over time, cause and effect. | At first, ____, but now ____. We saw that first ____, then ____, and finally ____. When I ____, it ____. After I ____, it ____. |
| **Notebook component—data organization** | | |
| Enumerate | Plot graphs: decide when and how to use bar graphs, line plots, and two-coordinate graphs to organize data. | Define x- and y-axes; provide axis labels, title, units, coordinates; use equal intervals, set origin = (0, 0). |
| Compare<br>Classify<br>Sequence | Use graphic organizers and narratives to express similarities and differences, to assign an object or action to the category or type to which it belongs, and to show sequencing and order. | This ____ is the same as ____ because ____. This ____ is different than ____ because ____. All these are ____ because ____. ____, ____, and ____ all have/are ____. |
| Analyze | Use graphic organizers, narratives or concept maps to identify part/whole or cause–and–effect relationships. | Use relationship verbs such as *contain, consist of*. As ____, then ____. When I changed ____, then ____ happened. The more/less ____, then ____. |

**Table 1 (*continued*)**

| Language function | Language objectives for writing in notebooks | Language forms, structures, and scaffolds for writing |
|---|---|---|
| **Notebook component—making sense of data** | | |
| Infer<br>Explain | Provide claims and evidence: write assertions about what was learned from the investigation, use the data as evidence to support those claims. | Use inferential logical connectors such as *although, while, thus, therefore*.<br>I claim that ____. I know this because ____. |
| Provide evidence | Use qualitative and quantitative data from the investigation as evidence to support claims. | Use qualitative descriptors such as *more/less, longer/shorter, brighter/dimmer*. Use quantitative expressions using standard metric units of measurement such as cm, mL, °C.<br>My data show ____. |
| Summarize<br>Predict<br>Generalize | Write a summary narrative to communicate what was learned; ask questions and make predictions based on the newly acquired knowledge. | Answer the focus question by rewriting it as a statement and providing evidence from data. Make a concluding statement. I learned ____. Therefore, I think ____. I predict ____ because ____.<br>A new question I have is ____. |
| **Notebook component—next-step strategies** | | |
| Critique<br>Evaluate | Reflect on experience: review notebook entries and revise, using line of learning or 3 Cs (correct in red, confirm in green, and complete in blue). | I used to think ____, but now I think ____. I have changed my thinking about ____.<br>I am confused about ____ because ____. I wonder ____. |

# READING DOMAIN

Reading is an integral part of science learning. Just as scientists spend a significant amount of their time reading one another's published works, students need to learn to read scientific text—to read effectively for understanding with a critical focus on the ideas being presented.

The articles in *FOSS Science Resources* help facilitate sense making as students make connections to the science concepts introduced and explored during the active investigations. Concept development is most effective when students are allowed to experience organisms, objects, and phenomena firsthand before engaging the concepts in text. The text and illustrations help students make connections between what they have experienced concretely and the abstract ideas that explain their observations.

*FOSS Science Resources* supports developing literacy skills by providing reading material that corresponds exactly to the concrete, personal experience provided in the active investigations. Students read with enthusiasm when they recognize familiar materials, organisms, and activities and are eager to tackle the reading to confirm their prior knowledge and discover more about the topic. In addition to making connections, once engaged, students naturally use other reading-comprehension strategies, such as asking questions, visualizing, inferring, and synthesizing, to help them understand the reading. As students apply these strategies, they are, in effect, using some of the same scientific thinking processes that promote critical thinking and problem solving.

## Reading in the Primary Grades

In the kindergarten modules, you can enhance science learning by using trade books and other read-aloud resources to engage students and provide topics for lively discussions. Reading aloud helps primary students understand the science content and lets you model reading-comprehension strategies, such as asking yourself questions (thinking aloud) and summarizing a paragraph just read. The Reading in *FOSS Science Resources* sections (in every part that has a reading) usually offer suggestions for activating prior knowledge before reading, indicate places to pause and discuss key points during the reading, and describe activities to deepen understanding after the reading. As students develop their reading skills, you might try these different ways to read from *FOSS Science Resources*.

- Read aloud from the big book while students follow along in their own books.

- Lead students in small guided reading groups.

- Have students read aloud with a partner.

- Have students read silently on their own.

The same strategies used in language arts can be applied to reading in science. Begin with reading the article aloud so that students can hear the content read fluently and listen for meaning and coherence. Go back and review the questions within the article. Use think–pair–share or other discussion protocols to allow students to think first, share with a partner, and then respond to the group. Emphasize blending phonemes as you read the text again, and model tracking, connecting spoken words with written words, and helping students identify sight words. Model reading-comprehension strategies by using think-alouds (as you think aloud, you explain the process that you are using in order to understand the text while reading). The expository text structure also provides the opportunity for primary students to learn how to extract information from a table of contents, a glossary, an index, and other text conventions such as headings, subheads, boldface and italic print, labeled graphics, and captions.

At the end of articles, use the questions provided to guide understanding and to assess comprehension and vocabulary acquisition. Choose one or two questions for students to answer in their notebooks. Emphasize the importance of science vocabulary and the appropriate language forms and structures. Here are some ways students can enhance reading comprehension.

- Have students predict the sequence of events or content.

- Have students write or dictate questions about the text and illustrations.

- Use visualization with students to "see, touch, feel, smell, hear" in their minds the content presented in the article.

- Ask students to make connections to information in the article and their observations during the active investigation.

# Reading in the Upper-Elementary Grades

As students progress from *learning to read* to *reading to learn*, they apply the strategies and skills of reading comprehension to learning about science in *FOSS Science Resources* and other texts. Writing becomes increasingly important in helping students make sense of the readings as well as the active investigations. Use the suggested questions in the *Investigations Guide* to support comprehension as students read from *FOSS Science Resources*. For most of the investigation parts, the articles are designed to follow the active investigation and are interspersed throughout the flow of the module. This allows students to acquire the necessary background knowledge through active experience before tackling the wider-ranging content and relationships presented in the text. Additional strategies for reading are derived from the seven essential strategies that readers use to help them understand what they read (Keene and Zimmermann 2007).

- Monitor for meaning: Discover when you know and when you don't know.

- Use and create schemata: Make connections between the novel and the known; activate and apply background knowledge.

- Ask questions: Generate questions before, during, and after reading that reach for deeper engagement with the text.

- Determine importance: Decide what matters most, what is worth remembering.

- Infer: Combine background knowledge with information from the text to predict, conclude, make judgments, and interpret.

- Use sensory and emotional images: Create mental images to deepen and stretch meaning.

- Synthesize: Create an evolution of meaning by combining understanding with knowledge from other texts/sources.

Following are some strategies that enhance the reading of expository texts in general and have proven to be particularly helpful in science.

**Build on background knowledge.** Activating prior knowledge is critical for helping students make connections between what they already know and new information. Reading comprehension improves when students have the opportunity to think, discuss, and write about what they know about a topic before reading. Review what students learned from the active investigation, provide prompts for making connections, and ask questions to help students recall past experiences and previous exposure to concepts related to the reading.

## Strategies for Reading in Upper-Elementary Grades

- *Build on background knowledge*
- *Create an anticipation guide*
- *Draw attention to vocabulary*
- *Preview the text*
- *Turn and talk*
- *Jigsaw text reading*
- *Note making*
- *Interactive reading aloud*
- *Summarize and synthesize*
- *3-2-1*
- *Write reflections*
- *Preview and predict*
- *SQ3R*

***Create an anticipation guide.*** Create true-or-false statements related to the key ideas in the reading selection. Ask students to indicate if they agree or disagree with each statement before reading, then have them read the text, looking for the information that supports their true-or-false claims. Anticipation guides connect students to prior knowledge, engage them with the topic, and encourage them to explore their own thinking.

***Draw attention to vocabulary.*** Check the article for bold words to determine if there are words students may not know. Review the science words that have already been listed on the word wall and in students' notebooks. For new science and nonscience vocabulary words that appear in the reading, have students predict their meanings before reading. During the reading, have students use strategies such as context clues and word structure to see if their predictions were correct. This strategy activates prior knowledge and engages students by encouraging analytical participation with the text.

***Preview the text.*** Give students time to skim through the selection, noting subheads, before reading thoroughly. Point out the particular structure of the text and what discourse markers to look for. For example, most *FOSS Science Resources* articles are written as cause and effect, problem and solution, question and answer, comparison and contrast, description, and sequence. Students will have an easier time making sense of the text if they know what to look for in terms of text structure. Model and have students practice analyzing these different types of expository text structures by looking for examples, patterns, and the discourse markers.

Point out how *FOSS Science Resources* text is organized (titles, headings, subheadings, questions, and summaries) and how to use the table of contents, glossary, and index. Explain how to scan for formatting features that provide key information (such as bold type and italics, captions, and framed text) and graphic features (such as tables, graphs, photographs, maps, diagrams, charts, and illustrations) that help clarify, elaborate, and explain important information in the reading.

While students preview the article, have them focus on the questions that appear in the text, as well as questions at the end of the article. Encourage students to write down questions they have that they think the reading will answer.

▶ **NOTE**
Discourse markers are words or phrases that relate one idea to another. Examples are "however," "on the other hand," and "second."

**Turn and talk.** When reading as a whole class, stop at key points and have students share their thinking about the selection with the student sitting next to them or in their collaborative group. This strategy helps students process the information and allows everyone to participate in the discussion. When reading in pairs, encourage students to stop and discuss with their partners. One way to encourage engagement and understanding during paired reading is to have students take turns reading aloud a paragraph or section on a certain topic. The one who is listening then summarizes the meaning conveyed in the passage.

**Jigsaw text reading.** Students work together in small groups (expert teams) to develop a collective understanding of a text. Each expert team is responsible for one portion of the assigned text. The teams read and discuss their portions to gain a solid understanding of the key concepts. They might use graphic organizers to refine and organize the information. Each expert team then presents its piece to the rest of the class. Or new, small jigsaw groups can be formed that consist of at least one representative from each expert team. Each student shares with the jigsaw group what their team learned from their particular portion of the text. Together, the participants in the jigsaw group fit their individual pieces together to create a complete picture of the content in the article.

**Note making.** The more interactive that students make a reading, the better for their understanding. Encourage students to become active readers by asking them to make notes as they read. Studies have shown that note making—especially paraphrasing and summarizing—is one of the most effective means for understanding text (Graham and Herbert 2010; Applebee 1984).

Students can annotate the text by writing thoughts and questions on self-stick notes. Using symbols or codes can help facilitate comprehension monitoring. Here are some possible symbols (Harvey 1998).

| | |
|---|---|
| ★ | interesting |
| BK | background knowledge |
| ? | question |
| C | confusing |
| I | important |
| L | learning something new |
| W | wondering |
| S | surprising |

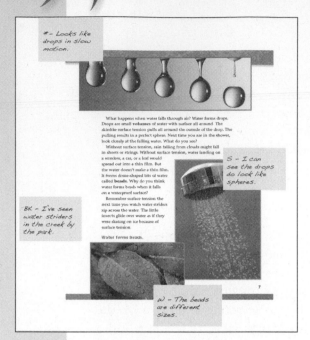

**An example of annotated text in *FOSS Science Resources***

Students can use a set of symbols while making notes about connections. The readings in *FOSS Science Resources* incorporate the active learning that students gain from the investigations so that they can make authentic text-to-self (T-S) connections. In other words, what they read reminds them of firsthand experiences, making the article more engaging and easier to understand. Text-to-text (T-T) connections are notes students make when they discover a new idea that reminds them of something they've read previously in another text. Text-to-world (T-W) connections involve the text and more global everyday connections to students' lives.

You can model note-making strategies by displaying a selection of text, using a projection system, a document camera, or an interactive whiteboard. As you read the text aloud, model how to write comments on self-stick notes, and use a graphic organizer in a notebook to enhance understanding.

Graphic organizers help students focus on extracting the important information from the reading and analyzing relationships between concepts. This can be done by simply having students make columns in their notebooks to record information and their thinking (Harvey and Goudvis 2007). Here are two examples of graphic organizers.

| Notes | Thinking |
| --- | --- |
| | |

| Facts | Questions | Responses |
| --- | --- | --- |
| | | |

***Interactive reading aloud.*** Reading aloud is an effective strategy for enhancing text comprehension. It offers opportunities to model specific reading-comprehension strategies and allows students to concentrate on making sense of the content. When modeling, share the thinking processes used to understand the reading (questioning, visualizing, comparing, inferring, summarizing, etc.), then have students share what they observed you thinking about as an active reader.

***Summarize and synthesize.*** Model how to pick out the important parts of the reading selection. Paraphrasing is one way to summarize. Have students record summaries of the reading, using their own words. To scaffold the learning, use graphic organizers to compare and contrast, group, sequence, and show cause and effect. Another method is to have students make two columns in their notebooks. In one column, they record what is important, and in the other they record their personal responses (what the reading makes them think about). When writing summaries, tell students,

- *Pick out the important ideas.*

- *Restate the main ideas in your own words.*

- *Keep it brief.*

***3-2-1.*** This strategy gives students the opportunity to synthesize information and formulate questions they still have regarding the concepts covered in an article. In their notebooks, students write three new things they learned, two interesting things worth remembering and sharing, and one question that occurred to them while reading the article. Other options might include three facts, two interesting ideas, and one insight about themselves as learners; three key words, two new ideas, and one thing to think about (modified from Black Hills Special Service Cooperative 2006).

***Write reflections.*** After reading, ask students to review their notes in their notebooks to make any additions, revisions, or corrections to what they recorded during the reading. This review can be facilitated by using a line of learning. Students draw a line under their original conclusion or under their answer to a question posed at the end of an article. They then add any new information as a new narrative entry. The line of learning indicates that what follows represents a change of thinking.

***Preview and predict.*** Instruct students to independently preview the article, directing attention to the illustrations, photos, boldfaced words, captions, and anything else that draws their attention. Working with a partner, students discuss and write three things they think they will learn from the article. Have partners verbally share their list with another pair of students. The group of four can collaborate to generate one list. Teams report their ideas, and together you create a class list on chart paper.

Read the article aloud, or have students read with a partner aloud or silently. Referring to the preview/prediction list, discuss what students learned. Have them record the most important thing they learned from the reading for comparison with the predictions.

***SQ3R.*** Survey! Question! Read! Recite! Review! provides an overall structure for before, during, and after reading. Students begin by previewing the text, looking for features that will help them make predictions about content. Based on their surveys, students develop questions to answer as they read. Students then read the selection. They recite by telling a partner what they've learned (their partner listens carefully to their recitation, making sure they haven't missed any important concepts) and answering any questions their partner might have. Finally, they review their questions and answers. There is value in asking students to make their entire SQ3R process explicit.

Use this script as an outline, using appropriate wording where necessary.

**Survey! Question!**

a. Before you read, survey (preview) the reading selection.

b. As you survey, think of questions that the reading might answer while you read the title, headings, and subheadings.

c. Read the questions at the end of the article.

d. Ask yourself, "What did my teacher say about this selection when it was assigned?"

e. Ask yourself, "What do I already know about this subject from experiments or discussions we've had in class?"

**Read!**

   a.  While reading, look for answers to the questions you raised.

   b.  Think about answers for the questions at the end of the article.

   c.  Study graphics, such as pictures, graphs, and tables.

   d.  Reread captions associated with pictures, graphs, and tables.

   e.  Note all italicized and boldfaced words or phrases.

   f.  Reduce your reading speed for difficult passages.

   g.  Stop and reread parts that are not clear.

   h.  Read only a section at a time, and recite after each section.

**Recite!**

   a.  After you've read a section, make notes in your science notebook by writing information in your own words.

   b.  Underline or highlight important notes in your science notebook.

   c.  Use the method of recitation that best suits your particular learning style. Remember that the more senses you use, the more likely you are to remember what you read. Seeing, saying, hearing, and writing will all enhance learning.

   d.  Ask questions aloud about what you just read, and summarize aloud, in your own words, what you read.

   e.  Listen attentively as your partner recites.

**Review!**

Reviewing should be ongoing. With the whole class, review the entire process, calling on groups to talk to the class about what they've learned.

## Struggling Readers

For students reading below grade level, the strategies listed above can be modified to support reading comprehension by integrating strategies from the Reading in the Primary Grades section, such as read-alouds and guided reading. Students can also follow along with the audio stories on FOSSweb. Breaking the reading down into smaller chunks, providing graphic organizers, and modeling reading-comprehension strategies can also help students who may be struggling with the text. For additional strategies for English learners, see the supported-reading strategy in the English-Language Development section of this chapter.

# SCIENCE-VOCABULARY DEVELOPMENT

Words play two critically important functions in science. First and most important, we play with ideas in our minds, using words. We present ourselves with propositions—possibilities, questions, potential relationships, implications for action, and so on. The process of sorting out these thoughts involves a lot of internal conversation, internal argument, weighing of options, and complex linguistic decisions. Once our minds are made up, communicating that decision, conclusion, or explanation in writing or through verbal discourse requires the same command of the vocabulary. Words represent intelligence; acquiring the precise vocabulary and the associated meanings is key to successful scientific thinking and communication.

The words introduced in FOSS investigations represent or relate to fundamental science concepts and should be taught in the context of the investigation. Many of the terms are abstract and are critical to developing science content knowledge and scientific and engineering practices. The goal is for students to use science vocabulary in ways that demonstrate understanding of the concepts the words represent—not to merely recite scripted definitions. The most effective science-vocabulary development strategies help students make connections to what they already know. These strategies focus on giving new words conceptual meaning through experience; distinguishing between informal, everyday language and academic language; and using the words in meaningful contexts.

## Building Conceptual Meaning through Experience

In most instances, students should be presented with new words in the context of the active experience at the need-to-know point in the investigation. Words such as *flexible*, *magnetic*, *viscous*, *silt*, *erode*, and *electricity* are conceptually loaded and significantly abstract. Students will have a much better chance of understanding, assimilating, and remembering the new word (or new meaning) if they can connect it with a concrete experience.

The new-word icon appears in the sidebar when you introduce a word that is critical to understanding the concepts or scientific practices students will be learning and applying in the investigation. When you introduce a new word, students should

- Hear it: Students listen as you model the correct contextual use and pronunciation of the word.

- See it: Students see the new word written out. Add a visual reference (an illustration or a sample) next to the word if possible.

- Say it: Have primary-grade students repeat the word chorally and clap out the syllables.

- Write it: You write the word on the board, chart paper, sentence strip, or card. Students use the new words in context when they write in their notebooks.

- Act it: Demonstrate action words such as *separate*, *compare*, and *observe* (using total physical response).

## Bridging Informal Language to Science Vocabulary

Students bring a wealth of language experience to the classroom. FOSS investigations are designed to tap into students' inquisitive natures and their excitement of discovery in order to encourage lively discussions as they explore materials in creative ways. There should be a lot of talking during science time! Your role is to help students connect informal language to the vocabulary used to express specific science concepts. As you circulate during active investigation, you continually model the use of science vocabulary. For example, when a student holds up a bottle of water and says, "I can see through it!" you might respond, "Yes, I see what you mean; the liquid in the bottle is transparent." Following are some strategies for validating students' conversational language while developing their familiarity with and appreciation for science vocabulary.

***Word bubbles.*** Choose a word from the word wall that is widely used by students and that is a synonym for a science vocabulary word. Draw a circle on the board or chart paper, and write the word in the center. Draw lines out from the circled word, and make more circles. Ask students to call out more synonyms for the word, and write them in the outer circles. Introduce the target vocabulary word as yet one more synonym for the target word, a word that is used in science. Highlight the word, and model its correct usage and pronunciation. Encourage students to use it in their discussions and in their notebook entries. Introduce the science word that is its opposite, when appropriate.

**Bridging Language Strategies**
- *Word bubbles*
- *Word sorts*
- *Semantic webs*
- *Concept maps*
- *Cognitive content dictionaries*
- *Word associations*

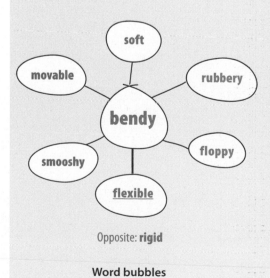

Opposite: **rigid**

**Word bubbles**

**Word sorts.** Make a set of word cards from words on the word wall. Ask students to help you group the words that are synonyms or that have conceptual connections. Add the new science words to the card set. Repeat the process with a new set of words. For upper-elementary students, make sets of vocabulary cards for students to sort in small groups.

**Semantic webs.** Select a vocabulary word, and write it in the center of a piece of paper (or on the board if doing this with the whole class). Brainstorm a list of words or ideas that are related to the first word. Group the words and concepts into several categories, and attach them to the central word with lines, forming a web (modified from Hamilton 2002).

**Concept maps.** Select six to ten related science words. Have students write them on self-stick notes or cards. Have small groups discuss how the words/concepts are related. Students organize words in groups and glue them down or copy them onto a sheet of paper. Students draw lines between the related words. On the lines, they write words describing or explaining how the concept words are related.

**Cognitive content dictionaries.** There are many variations of frameworks to help students record new words. This example can be used with the whole class with primary students or individually with upper-elementary students to introduce a few key vocabulary words used in an investigation. Have students write the word, predict its meaning, write the final meaning after class discussion (using primary language or an illustration when appropriate), and then use the word in a sentence. The word can then be used as a signal word to call for attention.

| Cognitive Content Dictionary | |
| --- | --- |
| New word | esker |
| Prediction (clues) | a flower growing at the North Pole |
| Final meaning (e.g., primary language, pictures) | a sand and gravel ridge, formed at the edges of a stream, flowing under a glacier |
| How I would use it (sentence) | The esker showed that a glacier had been in the valley at some time in the past. |

**Word associations.** In this brainstorming activity, you say a word, and students respond by writing the first word that comes to mind. Then students share their words with the class. This activity builds connections to students' prior frames of deference.

# Using Science Vocabulary in Context

In order for a new vocabulary word to become part of a student's functional vocabulary, he or she must have ample opportunities to hear and use it. The use of vocabulary terms is embedded in the activities through teacher talk, whole-class and small-group discussions during investigations, writing in science notebooks, readings, assessments, and games. In addition, other methods used during language-arts instruction can be used to reinforce important vocabulary words and phrases.

**Word wall/word cards.** Use chart paper or a pocket chart to record both science content and procedural words. Record the words as they come up during and after the investigations. Then copy key words on sentence strips or cards, and put them in a pocket chart. With a pocket chart, words can be sorted and moved around easily. For example, you could ask students to find words that are synonyms, antonyms, nouns, or verbs. Word cards should be available to each group during the investigation. This allows students to retrieve a word quickly when they are labeling diagrams and objects used during the investigation.

**Drawings and diagrams.** For English learners and visual learners, a diagram can be used to review and explain abstract content. Ahead of time, draw an illustration lightly, almost invisibly, with pencil on chart paper. When it's time for the investigation, trace the illustration with markers as you introduce the words and phrases to students. Students will be amazed by your artistic ability.

**Cloze activities.** Structure a sentence for students to complete, leaving out the vocabulary word, and crafting the sentence so that the missing vocabulary word is the last word in the sentence. You can do this chorally with primary students or in writing on the board or chart paper for upper-elementary students. Here's an example from the **Solids and Liquids Module**.

> Teacher: *Liquids that are clear and that you can see through are ____.*
>
> Students: *Transparent.*

## Science Vocabulary Strategies

- *Word wall/word cards*
- *Drawings and diagrams*
- *Cloze activities*
- *Word wizard*
- *Word analysis/word parts*
- *Breaking apart words*
- *Possible sentences*
- *Reading*
- *Highlighting the vocabulary*
- *Index*
- *Poems, chants, and songs*
- *Games*

**Word wizard.** Tell students that you are going to lead a word activity. You will be thinking of a science vocabulary word from the word wall. The goal is to figure out the word. Provide hints that have to do with parts of a definition, root word, prefix, suffix, and other relevant components. Students work in teams of two to four. Provide one hint, and give teams 1 minute to discuss. One team member writes the word on a piece of paper or on the whiteboard, using dark marking pens. Each team holds up its word for only you to see. After the third clue, reveal the word, and move on to the next word.

1. *This word is part of a plant.*

2. *It is usually not green.*

3. *It brings water and nutrients into the plant.*

*It is the* **root**.

**Word analysis/word parts.** Learning clusters of words that share a common origin can help students understand content-area texts and connect new words to familiar ones. This type of contextualized teaching meets the immediate need of understanding an unknown word while building generative knowledge that supports students in figuring out difficult words for future reading.

*geology*

*geologist*

*geological*

*geography*

*geometry*

*geophysical*

**Breaking apart words.** Have teams of two to four students break the word into prefix, root word, and suffix. Give each team different words, and have each team share the parsed elements of the word with the whole class.

**electromagnetism**

*electro*: having to do with electricity

*magnet*: having polar properties; attraction and repulsion of opposite and similar poles

*ism*: relating to a theory of how things behave

**Possible sentences.** Here is a simple strategy for teaching word meanings and generating class discussion.

1. Choose six to eight key concept words from the text of an article in *FOSS Science Resources*.

2. Choose four to six additional words that students are more likely to know something about.

3. Put the list of 10–14 words on the board or project it. Provide brief definitions as needed.

4. Ask students to devise sentences that include two or more words from the list.

5. On chart paper, write all sentences that students generate, both coherent and otherwise.

6. Have students read the article from which the words were extracted.

7. Revisit students' sentences, and discuss whether the sentences are sensible based on the passage or how they could be modified to be more coherent.

**Reading.** After the active investigation, students continue to develop their understanding of the vocabulary words and the concepts those words represent by listening to you read aloud, reading with a partner, or reading independently. Use strategies discussed in the Reading Domain section to encourage students to articulate their thoughts and practice the new vocabulary.

**Highlighting the vocabulary.** Emphasize the vocabulary words students should be using when they answer the focus question in their science notebooks. Distribute copies of the vocabulary/glossary for the investigation (available on FOSSweb) for students to glue into their notebooks. As you introduce the words in the investigation, students highlight them.

**Index.** Have students create an index at the back of their notebooks. There, they can record new vocabulary words and the notebook page where they defined and used the new words for the first time in the context of the investigation.

**Poems, chants, and songs.** Vocabulary words and phrases can be reinforced using content-rich poems, rhymes, chants, and songs.

**Games.** The informal activities included in the investigations are designed to reinforce important vocabulary words. Once students learn them, the words can be integrated into any type of independent work time, such as centers, workshops, and early-finisher tasks.

▶ **NOTE**
See the Science Notebooks in Grades 3–5 chapter for an example of this index.

▶ **NOTE**
English-language development refers to the advancement of students' ability to read, write, and speak English.

**EL  NOTE**

*Look for EL Notes in the investigations at points where students at beginning levels of English proficiency may need additional support.*

# ENGLISH-LANGUAGE DEVELOPMENT

Active investigations, together with ample opportunities to develop and use language, provide an optimal learning environment for English learners. This section highlights the English–language development (ELD) opportunities inherent in FOSS investigations and suggests other best practices for facilitating both the learning of new science concepts and the development of academic vocabulary and language structures that enhance literacy. For example, the hands–on structure of FOSS investigations is essential for the conceptual development of science content knowledge and the habits of mind that guide and define scientific practices. Students are engaged in concrete experiences that are meaningful and that provide a shared context for developing understanding—critical components for ELD instruction.

To further address the needs of English learners, the *Investigations Guide* includes EL Notes at points in the investigations where students at beginning levels of English proficiency may need additional support. When getting ready for an investigation, review the EL Notes, and determine the points where English learners may require scaffolds and where the whole class might benefit from additional language-development supports. One way to plan for ELD integration in science is to keep in mind four key areas: prior knowledge, comprehensible input, academic language development, and oral practice. The ELD chart below lists examples of universal strategies for each of these components that work particularly well in teaching science.

| English-Language Development (ELD) Quadrants | |
|---|---|
| **Activating prior knowledge**<br>• Inquiry chart<br>• Circle map<br>• Observation poster<br>• Quick write<br>• Kit inventory | **Using comprehensible input**<br>• Content objectives<br>• Multiple exposures<br>• Visual aids<br>• Supported reading<br>• Procedural vocabulary |
| **Developing academic language**<br>• Language objectives<br>• Sentence frames<br>• Word wall, word cards, drawings<br>• Concept maps<br>• Cognitive content dictionaries | **Providing oral practice**<br>• Small-group discussions<br>• Science talk<br>• Oral presentations<br>• Poems, chants, and songs<br>• Teacher feedback |

# Activating Prior Knowledge

When an investigation engages a new concept, first students recall and discuss familiar situations, objects, or experiences that relate to and establish a foundation for building new knowledge and conceptual understanding. Eliciting prior knowledge also supports learning by motivating interest, acknowledging culture and values, and checking for misconceptions and prerequisite knowledge. This is usually done in the first steps of Guiding the Investigation in the form of an oral discussion, presentation of new materials, or a written response to a prompt. The tools outlined below can also be used before beginning an investigation to establish a familiar context for launching into new material.

***Circle maps.*** Draw two concentric circles on chart paper. In the middle circle, write the topic to be explored. In the second circle, record what students already know about the subject. Ask students to think about how they know or learned what they already know about the topic. Record the responses outside the circles. Students can also do this independently in their science notebooks.

**Strategies for Activating Prior Knowledge**

- *Circle maps*
- *Observation posters*
- *Quick writes*
- *Kit inventories*

**An example of a circle map**

**Observation posters.** Make observation posters by gluing or taping pictures and artifacts relevant to the module or a particular investigation onto pieces of blank chart paper or poster paper. Hang them on the walls in the classroom, and have students rotate in small groups to each poster. At each station, students discuss their observations with their partners or small groups and then record (write, draw, or dictate) an observation, a question, a prediction, or an inference about the pictures as a contribution to the commentary on the poster.

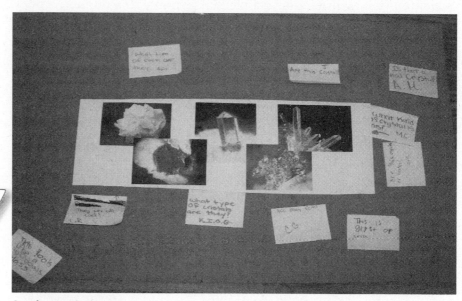

An observation poster

**Quick writes.** Ask students what they think they know about the topic of the investigation. Responses can be recorded independently as a quick write in science notebooks and then shared collaboratively. You should not correct misconceptions initially. Periodically revisit the quick-write ideas as a whole class, or have students review their notebook entries to correct, confirm, or complete their original thoughts as new information is acquired (possibly using a line of learning). At the conclusion of the investigation, students should be able to express mastery of the new conceptual material.

**Kit inventories.** Introduce each item from the FOSS kit used in the investigation, and ask students questions to get them thinking about what each item is and where they may have seen it before. Have them describe the objects and make predictions about how they will be used in the investigation. Tape samples of the items on chart paper, or print and display the equipment photo cards (download from FOSSweb) along with the name and a description, to serve as an interactive word wall.

Wood cylinder

| Item | Teacher | Student |
|------|---------|---------|
| Bottle | What is this? | A bottle. |
| | What is it made of? | Plastic. |
| | What is it used for? | To hold liquids. |
| Scoop | What is this? | It looks like a spoon. |
| | It's like a spoon. It's called a scoop. | Like an ice-cream scoop! |
| | What do you think we will be using it for in science? | To scoop things up. |
| Funnel | Have you seen this before? | My mom uses that for the car. |
| | It's called a funnel. Where else have you seen this? | In the kitchen. My uncle uses it sometimes to pour things. |
| | Can you describe it? | It's round on the ends. It's bigger on one end, and it's hollow. |
| Beaker | Have you seen this before? | We used it for science last year to measure and pour water. |
| | It's called a beaker. Where else have you seen this? | |
| Vial | This is a vial. What do you think we will be using it for in science? | To hold small things. |

**A kit inventory script from the Solids and Liquids Module**

# Comprehensible Input

In order to initiate their own sense-making process, students must be able to access the information presented to them. We refer to this ability as comprehensible input. Students must understand the essence of new ideas and concepts before beginning the process of constructing new scientific meaning. The strategies for comprehensible input used in FOSS ensure that the delivery of instruction is understandable while providing students with the opportunity to grapple with new ideas and the critically important relationships between concepts. Additional tools such as repetition, visual aids, emphasis on procedural vocabulary, and auditory reinforcement can also be used to enhance comprehensible input for English learners.

***Content objectives.*** The focus question for each investigation part frames the activity objectives—what students should know or be able to do at the end of the part. Making the learning objectives clear and explicit helps English learners prepare to process the delivery of new information, and helps you maintain the focus of the investigation. Write the focus question on the board, have students read it aloud and transcribe it into their science notebooks, and have students answer the focus question at the end of the investigation part. You then check their responses for understanding.

**Strategies for Comprehensible Input**
- *Content objectives*
- *Multiple exposures*
- *Visual aids*
- *Supported reading*
- *Procedural vocabulary*

**Procedural Vocabulary**

Add
Analyze
Assemble
Attach
Calculate
Change
Classify
Collect
Communicate
Compare
Connect
Construct
Contrast
Describe
Demonstrate
Determine
Draw
Evaluate
Examine
Explain
Explore
Fill
Graph
Identify
Illustrate
Immerse
Investigate
Label
List
Measure
Mix
Observe
Open
Order
Organize
Pour
Predict
Prepare
Record
Represent
Scratch
Separate
Sort
Stir
Subtract
Summarize
Test
Weigh

**Multiple exposures.** Repeat the activity as a class, at a center during independent work time, or in an analogous but slightly different context, ideally one that incorporates elements that are culturally relevant to students.

**Visual aids.** On the board or chart paper, write out the steps for conducting the investigation. This will provide a visual reference. Include illustrations if necessary. Use graphic representations (illustrations drawn and labeled in front of students) to review the concepts explored in the active investigations. In addition to the concrete objects included in the kit, use realia to augment the activity to help English learners build understanding and make cultural connections. Graphic organizers (webs, Venn diagrams, T-tables, flowcharts, etc.) aid comprehension by helping students see how ideas (concepts) are related.

**Supported reading.** In addition to the reading comprehension strategies suggested in the Reading Domain section of this chapter, English learners can also benefit from methods such as front-loading key words, phrases, and complex text structures before reading; using preview–review (main ideas are previewed in the primary language, read in English, and reviewed in the primary language); and having students use sentence frames specifically tailored to record key information and/or graphic organizers that make the content and the relationship between concepts visually explicit from the text as they read.

**Procedural vocabulary.** Make sure students understand the meaning of the words used in the directions describing what they should be doing during the investigation. These may or may not be science-specific words. Use techniques such as modeling, demonstrating, and body language (gestures) to explain procedural meaning in the context of the investigation. The words students will encounter in FOSS include those listed in the sidebar. To build academic literacy, English learners need to learn the multiple meanings of these words and their specific meanings in the context of science.

## Developing Academic Language

As students learn the nuances of the English language, it is critical that they build proficiency in academic language in order to participate fully in the cognitive demands of school. Academic language refers to the more abstract, complex, and specific aspects of language, such as the words, grammatical structure, and discourse markers that are needed for higher cognitive learning. FOSS investigations introduce and provide opportunities for students to practice using the academic vocabulary needed to access and meaningfully engage with science ideas.

**Language objectives.** Consider the language needs of English learners and incorporate specific language-development objectives that will support learning the science content of the investigation, such as a specific word knowledge skill (a way to expand use of vocabulary by looking at root words, prefixes, and suffixes), a linguistic pattern or structure for oral discussion and writing, or a reading-comprehension strategy. Recording in students' science notebooks is a productive place to optimize science learning and language objectives. For example, in the **Pebbles, Sand, and Silt Module**, one language objective might be "Students will describe what happened when they rubbed their rocks together (think-pair-share) and answer the focus question in their notebooks, using a cause–and–effect sentence frame."

**Vocabulary development.** The Science-Vocabulary Development section in this chapter describes the ways science vocabulary is introduced and developed in the context of an active investigation and suggests methods and strategies that can be used to support vocabulary development during instruction in English language arts and ELD. In addition to science vocabulary, students also need to learn the nonspecific-content words that facilitate deeper understanding and communication skills. Words such as *release, convert, beneficial, produce, receive, source,* and *reflect* are words used in the investigations and *FOSS Science Resources* and are frequently used in other content areas. Learning these academic vocabulary words gives students a more precise and complex way of practicing and communicating productive thinking. Consider using the strategies described in the Science-Vocabulary Development section to explicitly teach targeted, high-leverage words that can be used in multiple ways and that can help students make connections to other words and concepts. Sentence frames, word wall, concept maps, and cognitive content dictionary are strategies that have been found to be effective with academic-vocabulary development.

▶ **NOTE**

For additional resources and updated references, go to FOSSweb.

# REFERENCES

Applebee, A. 1984. "Writing and Reasoning." *Review of Educational Research* 54 (winter): 577–596.

Bereiter, C. 2002. *Education and Mind in the Knowledge Age.* Hillsdale, NJ: Erlbaum.

Black Hills Special Services Cooperative. 2006. "3-2-1 Strategy." In *On Target: More Strategies to Guide Learning.* Rapid City, SD. http://www.sdesa6.org/content/resources.htm.

Dyasi, H. M. 2006. "Visions of Inquiry: Science." In *Linking Science and Literacy in the K–8 Classroom,* ed. R. Douglas, K. Worth, and W. Binder. Arlington, VA: NSTA Press.

Gibbons, P. 2002. *Scaffolding Language, Scaffolding Learning.* Portsmouth, NH: Heinemann.

Graham, S., and M. Herbert. 2010. *Writing to Read: Evidence for How Writing Can Improve Reading.* New York: Carnegie.

Hamilton, G. 2002. *Content-Area Reading Strategies: Science.* Portland, ME: Walch Publishing.

Hand, B., and V. Prain. 2002. "Teachers Implementing Writing-to-Learn Strategies in Junior Secondary Science: A Case Study." *Science Education* 86: 737–755.

Harvey, S. 1998. *Nonfiction Matters: Reading, Writing, and Research in Grades 3–8.* Portland, ME: Stenhouse.

Harvey, S., and A. Goudvis. 2007. *Strategies That Work: Teaching Comprehension for Understanding and Engagement.* Portland, ME: Stenhouse.

Keene, E., and S. Zimmermann. 2007. *Mosaic of Thought: The Power of Comprehension Strategies.* 2nd ed. Portsmouth, NH: Heinemann.

Keys, C. 1999. *Revitalizing Instruction in Scientific Genres: Connecting Knowledge Production with Writing to Learn in Science.* Athens: University of Georgia.

Lieberman, G. A., and L.L. Hoody. 1998. *Closing the Achievement Gap: Using the Environment as an Integrating Context for Learning.* San Diego, CA: State Education and Environment Roundtable.

National Research Council. 2012. *A Framework for K–12 Science Education: Practices, Crosscutting Concepts, and Core Ideas.* Committee on Conceptual Framework for New Science Education Standards.

Ostlund, K. 1998. "What the Research Says about Science Process Skills: How Can Teaching Science Process Skills Improve Student Performance in Reading, Language Arts, and Mathematics?" *Electronic Journal of Science Education* 2 (4).

Wellington, J., and J. Osborne. 2001. *Language and Literacy in Science Education.* Buckingham, UK: Open University Press.

Winokur, J., and K. Worth. 2006. "Talk in the Science Classroom: Looking at What Students and Teachers Need to Know and Be Able to Do." In *Linking Science and Literacy in the K–8 Classroom*, ed. R. Douglas, K. Worth, and W. Binder. Arlington, VA: NSTA Press.

# FOSS and Common Core ELA – Grade 4

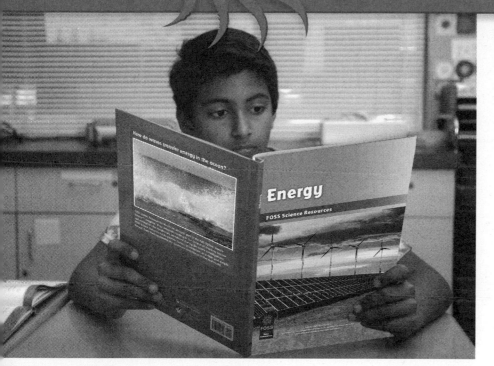

## Contents

Introduction .............................1

Reading Standards for
Informational Text ...................4

Reading Standards:
Foundational Skills...................8

Writing Standards.................. 10

Speaking and Listening
Standards.............................. 16

Language Standards............... 22

# INTRODUCTION

Each FOSS investigation follows a similar design to provide multiple exposures to science concepts. The design includes these pedagogies.

- Active investigation, including outdoor experiences

- Writing in science notebooks to answer focus questions

- Reading in *FOSS Science Resources*

- Assessment to monitor progress and motivate student reflection on learning

In practice, these components are seamlessly integrated into a continuum designed to maximize every student's opportunity to learn. An instructional sequence may move from one pedagogy to another and back again to ensure adequate coverage of a concept.

The FOSS instructional design recognizes the important role of language in science learning. Throughout the pedagogical design elements, students engage in the practices of the Common Core State Standards for English Language Arts. The purpose of this chapter is to provide the big picture of how FOSS provides opportunities for the development and exercising of these practices through science. The following pages have a table that identifies these opportunities in the three FOSS modules for the fourth grade.

## Guiding Principles

When integrating language-arts instruction with FOSS, keep in mind these guiding principles:

- FOSS investigations follow a clear and coherent conceptual flow and a consistent instructional design. Students develop science knowledge by building a framework of concepts and supporting ideas.

- Common Core State Standards for ELA are introduced, developed, and practiced in the context of learning science content and engaging in the science and engineering practices. Students read and comprehend complex science texts related to their prior experience and knowledge. They write informational/explanatory texts, arguments to support claims, and narratives about experiences in science. They engage in collaborative discussions about science and learn new vocabulary and language structures in context.

- The decision to use additional science texts, writing tasks, oral discourse opportunities, and vocabulary development activities is based on how well they address the science as well as the ELA standards.

- Instruction is differentiated to meet the needs of all students; the linguistic accommodations that are made for English learners support comprehensible input and accelerate academic language development. Language objectives for English learners in science instruction include the application of strategies that support construction of meaning from academic discussions and complex text, participation in productive discourse, and the ability to express ideas in writing clearly and coherently according to task, purpose, and audience.

- Formative assessment tools are used routinely to measure progress toward science understanding, use of science and engineering practices, and meeting literacy and language development goals. Assessment is viewed as a way to make student thinking visible and to determine next steps for instruction for both science and literacy. Instruction includes opportunities for students to assess themselves and peers.

Adhering to these guiding principles optimizes instructional time and, most importantly, benefits student learning by providing authentic and relevant contexts for building content knowledge, applying meaning-making strategies, and developing language and literacy skills.

Fourth grade marks an important transition to using literacy skills as tools for making meaning in content areas. Fourth graders read widely and deeply from a range of challenging informational texts that support their content learning and expand their vocabulary. They communicate in more complex and flexible ways that demonstrate understanding of purpose and audience. Fourth graders are expected to use the science and engineering practices to demonstrate their understanding of the core ideas. To accomplish this, students apply language structures for sequencing, comparing and contrasting, determining cause-and-effect relationships, and problem-solving.

## Instructional Flow

In almost all investigations, the instructional flow is the same and provides these opportunities for effective integration of ELA standards.

- When **setting the context** for the lesson, students activate prior knowledge through class or small-group discussions where they recount an experience using appropriate facts and relevant, descriptive details (SL 4).

- During the **active investigation**, students are expected to work with partners and in collaborative groups, and to engage in teacher-led discussions where they build on each other's ideas and express their own clearly (SL 1).

- In the **data management** phase, students make observations, and then routinely record and organize data in their notebooks (W 10). The notebook provides a space for students to recall information from experience, gather information from print and other media, take notes, and categorize information (W 8) and to acquire and use general academic and domain-specific words and phrases (L 6).

- The **analysis** phase involves discussing data, constructing and writing explanations, and engaging in argumentation. Here, students are making meaning by writing explanatory texts (W 2), writing opinion pieces supporting a point of view with reasons (W 1), or conducting short research projects that build knowledge through investigation of different aspects of a topic. (W 7).

- **Reading** articles in *FOSS Science Resources* and other recommended readings provides a plethora of opportunities to address all the fourth-grade reading standards for informational text.

- Lastly, the **assessment** tools and next-step strategies for engaging students in high-level critical thinking support the development of the Common Core State Standards capacities of the literate individual: demonstrate independence, build strong content knowledge, comprehend as well as critique, and value evidence.

Again, we have provided you with some examples of how FOSS connects to the fourth-grade ELA standards; there are many more opportunities waiting to be created and explored by you and your students.

> **TEACHING NOTE**
>
> *Throughout the fourth-grade FOSS modules, opportunities for addressing the ELA standards have been noted; however these examples should not be considered the only places for integrating literacy skills.*

# READING STANDARDS FOR INFORMATIONAL TEXT

| | Standard | Energy Module |
|---|---|---|
| **Key Ideas and Details** | 1. Refer to details and examples in a text when explaining what the text says explicitly and when drawing inferences from the text. | Discuss articles in *FOSS Science Resources* Inv 1, Part 1, Step 21; Inv 1, Part 2, Step 26 Inv 1, Part 3, Step 29; Inv 1, Part 4, Step 13 Inv 2, Part 2, Step 26; Inv 2, Part 3, Step 23 Inv 3, Part 1, Step 19; Inv 3, Part 3, Step 15 Inv 4, Part 2, Steps 17, 18; Inv 4, Part 3, Step 19 Inv 5, Part 1, Step 21; Inv 5, Part 2, Step 30; Inv 5, Part 3, Step 24 |
| | 2. Determine the main idea of a text and explain how it is supported by key details; summarize the text. | Discuss and review articles in *FOSS Science Resources* Inv 1, Part 2, Step 25; Inv 1, Part 3, Step 29 Inv 2, Part 2, Step 26; Inv 2, Part 3, Step 22 Inv 3, Part 1, Step 18; Inv 3, Part 2, Step 14 Inv 4, Part 1, Steps 15, 16; Inv 4, Part 3, Step 21 Inv 5, Part 2, Steps 24, 29, 30; Inv 5, Part 3, Step 23 |
| | 3. Explain events, procedures, ideas, or concepts in a historical, scientific, or technical text, including what happened and why, based on specific information in the text. | Discuss articles in *FOSS Science Resources* Inv 1, Part 1, Step 21; Inv 1, Part 2, Step 26 Inv 1, Part 3, Step 30; Inv 1, Part 4, Step 13 Inv 2, Part 2, Step 26; Inv 2, Part 3, Steps 19, 20, 23 Inv 3, Part 1, Step 19; Inv 3, Part 2, Step 16; Inv 3, Part 3, Step 15; Inv 4, Part 1, Step 15: Inv 4, Part 2, Step 19; Inv 4, Part 3, Step 19; Inv 5, Part 1, Step 21 Inv 5, Part 2, Steps 24, 30; Inv 5, Part 3, Step 24 |
| **Craft and Structure** | 4. Determine the meaning of general academic and domain-specific words or phrases in a text relevant to a *grade 4 topic or subject area*. | All investigations provide opportunities for students to determine the meaning of academic and science-specific words and phrases while reading articles in *FOSS Science Resources*. Selected examples Inv 1, Part 1, Step 21; Inv 1, Part 2, Step 25 Inv 2, Part 2, Step 25; Inv 3, Part 1, Step 19 Inv 4, Part 2, Steps 17, 18; Inv 4, Part 3, Steps 21-25 Inv 5, Part 1, Steps 19–20; Inv 5, Part 2, Step 2 |
| | 5. Describe the overall structure (e.g., chronology, comparison, cause/effect, problem/solution) of events, ideas, concepts, or information in a text or part of a text. | Discuss text structure in *FOSS Science Resources* Inv 1, Part 3, Step 30; Inv 1, Part 4, Step 12 Inv 2, Part 3, Step 18; Inv 2, Part 3, Steps 20, 21 Inv 3, Part 3, Step 14 Inv 4, Part 2, Step 18 Inv 5, Part 3, Step 25 |

*Common Core State Standards for English language arts and literacy in history/social studies science and technical subjects* (National Governors Association Center for Best Practices and Council of Chief State School Officers, 2010).

| Environments Module | Soils, Rocks, and Landforms Module |
|---|---|
| Discuss articles in *FOSS Science Resources*<br>Inv 1, Part 1, Steps 18, 33; Inv 1, Part 2, Steps 16, 28<br>Inv 1, Part 3, Step 14<br>Inv 2, Part 1, Step 23; Inv 2, Part 2, Steps 19, 20, 22<br>Inv 2, Part 3, Steps 22, 25; Inv 2, Part 4, Steps 18, 20<br>Inv 3, Part 1, Step 15; Inv 3, Part 2, Steps 9, 17<br>Inv 3, Part 3, Step 13; Inv 3, Part 4, Step 18<br>Inv 4, Part 1, Steps 45, 48, 50; Inv 4, Part 2, Step 17 | Discuss articles in *FOSS Science Resources*<br>Inv 1, Part 1, Step 20; Inv 1, Part 3, Step 20<br>Inv 2, Part 1, Step 20; Inv 2, Part 4, Steps 18, 21<br>Inv 3, Part 1, Step 22; Inv 3, Part 2, Step 14<br>Inv 3, Part 4, Step 11<br>Inv 4, Part 1, Steps 10, 12; Inv 4, Part 2, Step 11<br>Inv 4, Part 3, Step 16 |
| Discuss and review articles in *FOSS Science Resources*<br>Inv 1, Part 1, Steps 17, 18, 32, 33; Inv 1, Part 2, Steps 28, 29; Inv 1, Part 3, Step 13<br>Inv 2, Part 1, Step 21; Inv 2, Part 2, Steps 18, 20<br>Inv 2, Part 3, Step 21; Inv 2, Part 4, Steps 17-18, 20<br>Inv 3, Part 1, Step 15; Inv 3, Part 2, Steps 6, 17<br>Inv 3, Part 3, Step 13; Inv 3, Part 4, Step 17<br>Inv 4, Part 1, Step 45, 49; Inv 4, Part 2, Step 16 | Discuss and review articles in *FOSS Science Resources*<br>Inv 1, Part 1, Step 19<br>Inv 2, Part 1, Step 19; Inv 2, Part 4, Steps 17, 18<br>Inv 3, Part 1, Steps 21-22; Inv 3, Part 2, Step 14<br>Inv 3, Part 4, Steps 10, 11<br>Inv 4, Part 1, Step 9; Inv 4, Part 2, Step 11; Inv 4, Part 3, Steps 13, 16 |
| Discuss articles in *FOSS Science Resources*<br>Inv 1, Part 1, Steps 18, 33; Inv 1, Part 2, Steps 16, 29<br>Inv 2, Part 1, Step 21; Inv 2, Part 2, Steps 19, 20<br>nv 2, Part 3, Step 22; Inv 2, Part 4, Steps 18, 20<br>Inv 3, Part 2, Steps 9, 17; Inv 3, Part 4, Step 18<br>Inv 4, Part 1, Steps 45, 48; Inv 4, Part 2, Step 17 | Discuss articles in *FOSS Science Resources*<br>Inv 1, Part 3, Step 20<br>Inv 2, Part 1, Step 20; Inv 2, Part 4, Steps 17, 18, 22<br>Inv 3, Part 1, Step 22; Inv 3, Part 4, Steps 10, 11<br>Inv 4, Part 3, Step 16 |
| All investigations provide opportunities for students to determine the meaning of academic and science-specific words and phrases while reading articles in *FOSS Science Resources*.<br>Selected examples<br>Inv 1, Part 1, Steps 17, 18; Inv 1, Part 2, Step 16<br>Inv 2, Part 2, Step 18; Inv 2, Part 4, Steps 16-20<br>Inv 3, Part 1, Step 15; Inv 3, Part 2, Steps 9, 17<br>Inv 4, Part 1, Step 50; Inv 4, Part 2, Step 17 | All investigations provide opportunities for students to determine the meaning of academic and science-specific words and phrases while reading articles in *FOSS Science Resources*.<br>Selected examples<br>Inv 1, Part 1, Step 19; Inv 1, Part 3, Step 20; Inv 1, Part 3, Step 20; Inv 2, Part 4, Step 22; Inv 3, Part 2, Step 13<br>Inv 4, Part 1, Step 11; Inv 4, Part 2, Step 11; Inv 4, Part 3, Step 16 |
| Discuss text structure in *FOSS Science Resources*<br>Inv 1, Part 1, Step 16; Inv 1, Part 2, Step 16<br>Inv 1, Part 3, Step 12<br>Inv 2, Part 1, Step 20; Inv 2, Part 2, Steps 3, 4<br>Inv 2, Part 3, Steps 21, 22; Inv 2, Part 4, Step 20<br>Inv 3, Part 2, Step 16; Inv 3, Part 3, Step 12 | Discuss text structure in *FOSS Science Resources*<br>Inv 1, Part 3, Step 20<br>Inv 3, Part 2, Step 13<br>Inv 4, Part 3, Step 14 |

# READING STANDARDS FOR INFORMATIONAL TEXT

| | Standard | Energy Module |
|---|---|---|
| **Craft and Structure** | 6. Compare and contrast a firsthand and secondhand account of the same event or topic; describe the differences in focus and the information provided. | Read and compare articles in *FOSS Science Resources* to the active investigation.<br>Selected examples<br>Inv 1, Part 1, Step 21<br>Inv 2, Part 2, Step 25<br>Inv 3, Part 1, Step 17;  Inv 3, Part 3, Step 15<br>Inv 4, Part 1, Steps 14, 15; Inv 4, Part 2, Step 19<br>Inv 4, Part 3, Step 17<br>Inv 5, Part 1, Step 20 |
| **Integration of Knowledge and Ideas** | 7. Interpret information presented visually, orally, or quantitatively (e.g., in charts, graphs, diagrams, time lines, animations, or interactive elements on Web pages) and explain how the information contributes to an understanding of the text in which it appears. | Analyze and discuss illustrations used to support understanding in *FOSS Science Resources*.<br>Selected examples<br>Inv 1, Part 1, Step 19; Inv 1, Part 2, Step 24<br>Inv 1, Part 3, Step 29; Inv 1, Part 4, Step 12<br>Inv 2, Part 2, Step 25; Inv 2, Part 3, Steps 21, 22<br>Inv 3, Part 1, Step 17; Inv 3, Part 2, Step 16<br>Inv 4, Part 1, Steps 13, 15; Inv 4, Part 2, Steps 17, 18<br>Inv 4, Part 3, Steps 17, 18, 21, 24<br>Inv 5, Part 3, Step 25 |
| | 8. Explain how an author uses reasons and evidence to support particular points in a text. | Read and discuss articles in *FOSS Science Resources*<br>Inv 1, Part 3, Step 29<br>Inv 2, Part 3, Steps 21, 22<br>Inv 4, Part 3, Step 22 |
| | 9. Integrate information from two texts on the same topic in order to write or speak about the subject knowledgeably. | Students can read their *FOSS Science Resources* as well as readings suggested on FOSSweb. Using these two texts, students integrate the information when speaking and writing about science content.<br>Selected examples<br>Inv 1, Language Extensions. Read Mr. Henshaw<br>Inv 3, Part 3, Step 16<br>Inv 4, Part 3, Step 22<br>Inv 5, Language Extensions. Travel through the looking glass. |
| **Range of Reading** | 10. By the end of year, read and comprehend informational texts, including history/social studies, science, and technical texts, in the grades 4–5 text complexity band proficiently, with scaffolding as needed at the high end of the range. | All investigations provide opportunities for students to develop their ability to read and comprehend complex informational science text such as *FOSS Science Resources*. |

| Environments Module | Soils, Rocks, and Landforms Module |
|---|---|
| Read and compare articles in *FOSS Science Resources* to the active investigation.<br>Selected examples<br>Inv 1, Part 2, Steps 27, 29; Inv 1, Part 3, Step 14<br>Inv 2, Part 4, Step 18<br>Inv 3, Part 1, Step 14; Inv 3, Part 2, Steps 9, 15, 17<br>Inv 4, Part 1, Step 47 | Read and compare articles in *FOSS Science Resources* to the active investigation.<br>Selected examples<br>Inv 1, Part 1, Step 19<br>Inv 2, Part 1, Step 19; Inv 2, Part 4, Step 18<br>Inv 3, Part 1, Step 20 |
| Analyze and discuss illustrations used to support understanding in *FOSS Science Resources*.<br>Selected examples<br>Inv 1, Part 1, Steps 16-18, 31; Inv 1, Part 2, Steps 14, 27<br>Inv 2, Part 1, Steps 20, 21, Inv 2, Part 2, Steps 18, 19, 21<br>Inv 2, Part 3, Step 21; Inv 2, Part 4, Steps 16, 19<br>Inv 3, Part 1, Steps 14-15; Inv 3, Part 2, Steps 7, 15<br>Inv 3, Part 4, Step 17<br>Inv 4, Part 1, Steps 47, 49; Inv 4, Part 2, Step 16 | Analyze and discuss illustrations used to support understanding in *FOSS Science Resources*.<br>Selected examples<br>Inv 1, Part 1, Step 19; Inv 1, Part 3, Steps 19, 20<br>Inv 2, Part 1, Step 18; Inv 2, Part 4, Steps 17, 20, 21<br>Inv 3, Part 1, Step 20; Inv 3, Part 2, Step 14<br>Inv 3, Part 4, Step 10<br>Inv 4, Part 1, Steps 9, 13; Inv 4, Part 2, Step 10<br>Inv 4, Part 3, Steps 12, 14 |
| Read and discuss articles in *FOSS Science Resources*<br>Inv 1, Part 1, Step 33<br>Inv 2, Part 3, Step 24; Inv 2, Part 4, Step 20<br>Inv 4, Part 1, Step 49 | Read and discuss articles in *FOSS Science Resources*<br>Inv 1, Part 1, Steps 19, 20; Inv 1, Part 3, Step 19<br>Inv 2, Part 4, Step 20 |
| Students can read *FOSS Science Resources* as well as readings suggested on FOSSweb. Using these two texts, students integrate the information when speaking and writing about science content.<br>Selected examples<br>Inv 1, Part 3, Step 15<br>Inv 2, Part 2, Step 24; Inv 2, Part 3, Steps 15, 23<br>Inv 3, Part 3, Step 15<br>Inv 4, Part 1, Step 49; Inv 4, Part 2, Step 18 | Students can read *FOSS Science Resources* as well as readings suggested on FOSSweb. Using these two texts, students integrate the information when speaking and writing about science content.<br>Selected example<br>Inv 3, Part 2, Step 15 |
| All investigations provide opportunities for students to develop their ability to read and comprehend complex informational science text such as *FOSS Science Resources*. | All investigations provide opportunities for students to develop their ability to read and comprehend complex informational science text such as *FOSS Science Resources*. |

# READING STANDARDS: FOUNDATIONAL SKILLS

| | Standard | Energy Module |
|---|---|---|
| **Phonics and Word Recognition** | 3. Know and apply grade-level phonics and word analysis skills in decoding words.<br>  a. Use combined knowledge of all letter-sound correspondences, syllabication patterns, and morphology (e.g., roots and affixes) to read accurately unfamiliar multisyllabic words in context and out of context. | All investigations provide opportunities for students to apply decoding skills while reading articles in *FOSS Science Resources*.<br><br>Selected example<br>Inv 1, Part 1, Step 21 |
| **Fluency** | 4. Read with sufficient accuracy and fluency to support comprehension.<br>  a. Read grade-level text with purpose and understanding.<br>  b. Read grade-level prose and poetry orally with accuracy, appropriate rate, and expression on successive readings.<br>  c. Use context to confirm or self-correct word recognition and understanding, rereading as necessary. | All investigations provide opportunities for students to practice reading with accuracy and fluency.<br><br>Selected examples<br>Inv 1, Part 1, Step 20; Inv 1, Part 2, Step 25<br>Inv 1, Part 3, Step 29; Inv 1, Part 4, Step 12<br>Inv 2, Part 2, Step 25; Inv 2, Part 3, Steps 19, 22<br>Inv 3, Part 1, Step 19; Inv 3, Part 2, Steps 14, 16<br>Inv 3, Part 3, Step 14<br>Inv 4, Part 1, Step 14; Inv 4, Part 2, Step 18<br>Inv 4, Part 3, Steps 18, 21, 24<br>Inv 5, Part 1, Step 19; Inv 5, Part 2, Steps 23, 26, 29<br>Inv 5, Part 3, Steps 23, 25 |

| Environments Module | Soils, Rocks, and Landforms Module |
|---|---|
| All investigations provide opportunities for students to apply decoding skills while reading articles in *FOSS Science Resources*.<br><br>Selected examples<br>Inv 1, Part 1, Steps 17, 18; Inv 1, Part 2, Step 16<br>Inv 2, Part 4, Step 18 | All investigations provide opportunities for students to apply decoding skills while reading articles in *FOSS Science Resources*.<br><br>Selected examples<br>Inv 1, Part 1, Step 19; Inv 1, Part 3, Step 20<br>Inv 2, Part 4, Step 22 |
| All investigations provide opportunities for students to practice reading with accuracy and fluency.<br><br>Selected examples<br>Inv 1, Part 1, Step 17; Inv 1, Part 2, Steps 15, 28<br>Inv 1, Part 3, Steps 12, 13<br>Inv 2, Part 1, Step 21, 22; Inv 2, Part 2, Step 18<br>Inv 2, Part 3, Steps 21, 24; Inv 2, Part 4, Steps 17, 19<br>Inv 3, Part 1, Step 15; Inv 3, Part 2, Steps 6, 15, 16<br>Inv 3, Part 3, Step 12; Inv 3, Part 4, Step 17<br>Inv 4, Part 1, Steps 45, 47, 49; Inv 4, Part 2, Step 16 | All investigations provide opportunities for students to practice reading with accuracy and fluency.<br><br>Selected examples<br>Inv 1, Part 1, Step 19<br>Inv 2, Part 1, Step 19; Inv 2, Part 4, Steps 17, 21<br>Inv 3, Part 1, Step 21; Inv 3, Part 2, Step 13<br>Inv 3, Part 4, Step 10<br>Inv 4, Part 1, Step 9; Inv 4, Part 2, Step 10 |

## WRITING STANDARDS

| Standard | Energy Module |
|---|---|
| **Text Types and Purposes** | |
| 1. Write opinion pieces on topics or texts, supporting a point of view with reasons and information.<br>  a Introduce a topic or text clearly, state an opinion, and create an organizational structure in which related ideas are grouped to support the writer's purpose.<br>  b. Provide reasons that are supported by facts and details.<br>  c. Link opinion and reasons using words and phrases (e.g., *for instance, in order to, in addition*).<br>  d Provide a concluding statement or section related to the opinion presented. | All investigations provide opportunities for students to write their opinion, or claim, supported by reasons. Students answer questions (focus question, response sheets, assessments) by stating their claim supported by evidence and reasoning.<br><br>Selected example<br>Inv 1, Part 4, Step 10 |
| 2. Write informative/explanatory texts to examine a topic and convey ideas and information clearly.<br>  a. Introduce a topic clearly and group related information in paragraphs and sections; include formatting (e.g., headings), illustrations, and multimedia when useful to aiding comprehension.<br>  b. Develop the topic with facts, definitions, concrete details, quotations, or other information and examples related to the topic.<br>  c. Link ideas within categories of information using words and phrases (e.g., *another, for example, also, because*).<br>  d. Use precise language and domain-specific vocabulary to inform about or explain the topic.<br>  e. Provide a concluding statement or section related to the information or explanation presented. | All investigations provide opportunities for students to write explanatory texts to examine the science topic they are learning. In every part, students write an explanation as part of their answer to the focus question or the response sheet.<br><br>Selected examples<br>Inv 1, Part 1, Step 16; Inv 1, Part 2, Step 22<br>Inv 1, Part 3, Steps 14, 26, 27, 30<br>Inv 2, Part 1, Step 17; Inv 2, Part 2, Steps 22, 23<br>Inv 2, Part 3, Step 17<br>Inv 3, Part 1, Step 15; Inv 3, Part 2, Step 11<br>Inv 4, Part 2, Step 19; Inv 4, Part 3, Step 15<br>Inv 5, Part 2, Step 20<br><br>Students write a summary paragraph in the Wrap-up section at the end of every investigation. |
| 3. Write narratives to develop real or imagined experiences or events using effective technique, descriptive details, and clear event sequences.<br>  a. Orient the reader by establishing a situation and introducing a narrator and/or characters; organize an event sequence that unfolds naturally.<br>  b. Use dialogue and description to develop experiences and events or show the responses of characters to situations.<br>  c. Use a variety of transitional words and phrases to manage the sequence of events.<br>  d. Use concrete words and phrases and sensory details to convey experiences and events precisely.<br>  e. Provide a conclusion that follows from the narrated experiences or events. | All investigations provide opportunities for students to write narratives. Students describe their observations and experiences with the science ideas they are exploring.<br><br>Selected examples<br>Inv 1, Language Extensions. Imagine no electricity<br>Inv 2, Language Extensions. Write a story<br>Inv 3, Part 3, Step 12 |

| Environments Module | Soils, Rocks, and Landforms Module |
|---|---|
| All investigations provide opportunities for students to write their opinion, or claim, supported by reasons. Students answer questions (focus question, response sheets, assessments) by stating their claim supported by evidence and reasoning.<br><br>Selected examples<br>Inv 1, Part 2, Steps 19, 33, 36, 37<br>Inv 2, Part 2, Step 22; Inv 2, Part 4, Step 15<br>Inv 3, Part 3, Step 8<br>Inv 4, Part 2, Step 15 | All investigations provide opportunities for students to write their opinion, or claim, supported by reasons. Students answer questions (focus question, response sheets, assessments) by stating their claim supported by evidence and reasoning.<br><br>Selected examples<br>Inv 1, Part 2, Step 14<br>Inv 3, Part 3, Step 13; Inv 3, Part 4, Step 8<br>Inv 4, Part 1, Steps 7, 8; Inv 4, Part 3, Step 11 |
| All investigations provide opportunities for students to write explanatory texts to examine the science topic they are learning. In every part, students write an explanation as part of their answer to the focus question or the response sheet.<br><br>Selected examples<br>Inv 1, Part 1, Steps 18, 21; Inv 1, Part 3, Step 10<br>Inv 1, Language Extension, Write organism booklets<br>Inv 2, Part 1, Step 9; Inv 2, Part 2, Steps 12, 25<br>Inv 2, Part 3, Step 19; Inv 2, Part 4, Steps 15, 18<br>Inv 3, Part 3, Step 8<br>Inv 4, Part 2, Step 14<br><br>Students write a summary paragraph in the Wrap-up section at the end of every investigation. | All investigations provide opportunities for students to write explanatory texts to examine the science topic they are learning. In every part, students write an explanation as part of their answer to the focus question or the response sheet.<br><br>Selected examples<br>Inv 1, Part 2, Step 13; Inv 1, Part 3, Step 22<br>Inv 2, Part 1, Step 14; Inv 2, Part 2, Steps 7, 15; Part 4, Step 14<br>Inv 2, Language Extension, Write an investigation report<br>Inv 3, Part 1, Step 18; Inv 3, Part 2, Steps 10-11<br>Inv 3, Part 4, Step 6<br><br>Students write a summary paragraph in the Wrap-up section at the end of every investigation. |
| All investigations provide opportunities for students to write narratives. Students describe their observations and experiences with the science ideas they are exploring.<br><br>Selected example<br>Inv 3, Part 1, Step 13 | All investigations provide opportunities for students to write narratives. Students describe their observations and experiences with the science ideas they are exploring.<br><br>Selected examples<br>Inv 1, Part 3, Steps 4, 5<br>Inv 1, Language Extension. Write Soil Stories<br>Inv 2, Part 2, Step 23; Inv 2, Part 3, Step 10<br>Inv 3, Language Extensions. Create maps of fictional places |

## WRITING STANDARDS (*continued*)

| Standard | Energy Module |
|---|---|
| **Production and Distribution of Writing** | |
| 4. Produce clear and coherent writing in which the development and organization are appropriate to task, purpose, and audience. (Grade-specific expectations for writing types are defined in standards 1–3.) | All investigations provide opportunities for students to record and organize their data in their science notebooks. Based on their data, students construct and write their explanations.<br><br>Selected examples<br><br>Inv 1, Part 1, Step 16; Inv 1, Part 2, Steps 22, 24; Inv 1, Part 3, Steps 11, 14, 30; Inv 1, Part 4, Step 10<br>Inv 2, Part 1, Step 17; Inv 2, Part 2, Steps 20, 22, 23; Inv 2, Part 3, Step 17; Inv 2, Language Extensions. Write directions for compass use<br>Inv 3, Part 2, Step 11; Inv 3, Part 3, Step 12<br>Inv 4, Part 2, Step 14; Inv 4, Part 2, Step 19 |
| 5. With guidance and support from peers and adults, develop and strengthen writing as needed by planning, revising, and editing. (Editing for conventions should demonstrate command of Language standards 1–3 up to and including grade 4.) | The Wrap-up/Warm-up section of each investigation part provides the opportunity for students to strengthen their notebook entries by revising and adding in new information.<br><br>Inv 3, Part 1, Step 13<br>Inv 4, Part 2, Step 19<br><br><br>The Wrap-up review focus question section at the end of each investigation and next-step strategies after answering the response sheets or taking the I-Check also serve as a method for strengthening writing. |
| 6. With some guidance and support from adults, use technology, including the Internet, to produce and publish writing as well as to interact and collaborate with others; demonstrate sufficient command of keyboarding skills to type a minimum of one page in a single sitting. | |

| Environments Module | Soils, Rocks, and Landforms Module |
|---|---|
| All investigations provide opportunities for students to record and organize their data in their science notebooks. Based on their data, students construct and write their explanations.<br><br>Selected examples<br>Inv 1, Part 1, Steps 18, 21; Inv 1, Part 2, Steps 19, 33; Inv 1, Part 3, Step 10<br>Inv 2, Part 1, Step 9; Inv 2, Part 2, Steps 12, 19; Inv 2, Part 4, Steps 14, 15, 18<br>Inv 3, Part 1, Step 13; Inv 3, Part 2, Step 13; Inv 3, Part 3, Steps 8, 10<br>Inv 4, Part 2, Steps 14, 15 | All investigations provide opportunities for students to record and organize their data in their science notebooks. Based on their data, students construct and write their explanations.<br><br>Selected examples<br>Inv 1, Part 1, Step 16; Inv 1, Part 2, Steps 6, 8<br>Inv 2, Part 1, Step 14; Inv 2, Part 2, Steps 7, 13, 15, 20-21, 23; Inv 2, Part 3, Steps 4, 10; Inv 2, Part 3, Step 11; Inv 2, Part 4, Step 14<br>Inv 3, Part 1, Steps 18, 20; Inv 3, Part 2, Step 10; Inv 3, Part 3, Step 13; Inv 3, Part 4, Step 6<br>Inv 4, Part 1, Steps 7, 8; Inv 4, Part 3, Step 11 |
| The Wrap-up/Warm-up section of each investigation part provides the opportunity for students to strengthen their notebook entries by revising and adding in new information.<br><br>Inv 1, Part 1, Step 34<br>Inv 2, Part 1, Step 12; Inv 2, Part 3, Step 22; Inv 2, Part 4, Step 18<br>Inv 3, Part 2, Step 31<br>Inv 4, Part 1, Step 48<br><br>The Wrap-up review focus question section at the end of each investigation and next-step strategies after answering the response sheets or taking the I-Check also serve as a method for strengthening writing. | The Wrap-up/Warm-up section of each investigation part provides the opportunity for students to strengthen their notebook entries by revising and adding in new information.<br><br>Inv 1, Part 1, Step 21<br>Inv 2, Part 1, Step 21; Inv 2, Part 2, Steps 20, 28; Inv 2, Part 4, Step 19<br><br>The Wrap-up review focus question section at the end of each investigation and next-step strategies after answering the response sheets or taking the I-Check also serve as a method for strengthening writing. |
| | |

## WRITING STANDARDS (*continued*)

| Standard | Energy Module |
|---|---|
| **7.** Conduct short research projects that build knowledge through investigation of different aspects of a topic. | All investigations provide opportunities for students to further investigate different aspects of the science topic.<br><br>Inv 1, Part 1, Step 21; Inv 1, Part 2, Step 12; Inv 1, Part 3, Steps 21, 22, 31; Inv 1, Language Extensions. Research solar cell technology<br>Inv 2, Science and Engineering Extensions. Conduct more force investigations<br>Inv 3, Language Extensions. Research forms of code. Investigate emergency codes<br>Inv 4, Language Extensions. Research safety technologies. Research roller coasters |
| **8.** Recall relevant information from experiences or gather relevant information from print and digital sources; take notes and categorize information, and provide a list of sources. | All investigations provide students with the opportunity to write about their science experiences and record their observations in their science notebooks.<br>Students also take notes and organize information when reading articles in *FOSS Science Resources*.<br><br>Selected examples<br>Inv 1, Part 1, Steps 15, 16, 20; Inv 1, Part 2, Steps 12, 16, 22, 24; Inv 1, Part 3, Steps 11, 14, 20, 22, 23, 26-28, 30<br>Inv 1, Part 4, Steps 6, 10<br>Inv 2, Part 1, Steps 11, 15, 17, 21, 25<br>Inv 2, Part 2, Steps 9, 11, 13, 20, 22, 23, 27<br>Inv 2, Part 3, Steps 6, 12, 14, 16, 21<br>Inv 3, Part 1, Steps 8, 10; Inv 3, Part 2, Steps 10, 11, 14<br>Inv 3, Part 3, Step 14<br>Inv 4, Part 1, Steps 5, 12, 14, 17; Inv 4, Part 2, Steps 14, 19<br>Inv 4, Part 3, Steps 8, 15, 25<br>Inv 5, Part 1, Step 17; Inv 5, Part 2, Step 20<br>Inv 5, Part 3, Step 21 |
| **9.** Draw evidence from literary or informational texts to support analysis, reflection, and research.<br>a. Apply *grade 4 Reading standards* to literature (e.g., "Describe in depth a character, setting, or event in a story or drama, drawing on specific details in the text [e.g., a character's thoughts, words, or actions].").<br>b. Apply *grade 4 Reading standards* to informational texts (e.g., "Explain how an author uses reasons and evidence to support particular points in a text"). | All investigations provide opportunities to use the *FOSS Science Resources* as a source from which to draw evidence to support their ideas (e.g., discussion questions at the end of the articles).<br><br>Selected examples<br>Inv 1, Part 3, Step 30<br>Inv 2, Part 3, Step 23<br>Inv 3, Part 3, Step 15<br>Inv 4, Part 2, Step 19 |

Rows 7 and 8 are under the side label **Research to Build and Present Knowledge**. Row 9 is under the side label **Range of Writing**.

| **Environments Module** | **Soils, Rocks, and Landforms Module** |
|---|---|
| All investigations provide opportunities for students to further investigate different aspects of the science topic.<br><br>Inv 1, Part 1, Steps 11, 12, 26; Inv 1, Part 2, Steps 6, 11, 17-19, 20-26, 30-33; Science and Engineering Extensions, Research pheromones<br>Inv 2, Part 4, Step 18; Inv 2, Language Extension, Describe aquatic environments<br>Inv 3, Part 2, Steps 20-22; Inv 3, Science Extension, Observe the life cycle of brine shrimp<br>Inv 4, Part 1, Steps 9-12 | All investigations provide opportunities for students to further investigate different aspects of the science topic.<br><br>Inv 2, Part 2, Steps 13, 20-21<br>Inv 2, Part 3, Step 4;<br>Inv 3, Science Extension, Find out what surveyors do<br>Inv 4, Part 2, Step 11; Inv 4, Language Extension, Research natural resources in your region |
| All investigations provide students with the opportunity to write about their science experiences and record their observations in their science notebooks.<br>Students also take notes and organize information when reading articles in *FOSS Science Resources*.<br><br>Selected examples<br>Inv 1, Part 1, Steps 4, 12, 17, 19, 21; Inv 1, Part 2, Steps 9, 11, 15, 19, 28, 30, 33, 34; Inv 1, Part 3, Step 7; Inv 1, Part 3, Step 10<br>Inv 2, Part 1, Steps 1, 7, 9, 14, 20-23; Inv 2, Part 2, Step 12<br>Inv 2, Part 2, Steps 12, 18, 20, 22<br>Inv 2, Part 3, Steps 16, 17, 19; Inv 2, Part 3, Step 21<br>Inv 2, Part 4, Steps 14, 15, 18-20<br>Inv 3, Part 1, Steps 13, 15; Inv 3, Part 2, Steps 3, 5, 9, 16, 22, 26; Inv 3, Part 3, Step 8; Inv 3, Part 4, Step 16<br>Inv 4, Part 1, Steps 22, 28, 29, 32, 43, 47, 48<br>Inv 4, Part 2, Steps 14, 15; Inv 4, Part 3, Steps 3, 6 | All investigations provide students with the opportunity to write about their science experiences and record their observations in their science notebooks.<br>Students also take notes and organize information when reading articles in *FOSS Science Resources*.<br><br>Selected examples<br>Inv 1, Part 1, Step 16; Inv 1, Part 2, Steps 6, 8<br>Inv 1, Part 3, Step 15; Inv 1, Part 4, Steps 9, 13, 15<br>Inv 2, Part 1, Steps 5, 14; Inv 2, Part 2, Steps 7, 13, 15, 20-21, 23; Inv 2, Part 3, Steps 4, 10; Part 4, Steps 14, 16, 20, 21<br>Inv 3, Part 1, Steps 18, 21; Inv 3, Part 2, Steps 10, 13, 14<br>Inv 3, Part 3, Steps 12, 13; Inv 3, Part 4, Steps 3, 4<br>Inv 3, Part 4, Steps 9, 11<br>Inv 4, Part 1, Step 2; Inv 4, Part 1, Steps 10, 1<br>Inv 4, Part 2, Step 10; Inv 4, Part 3, Steps 3-5, 8 |
| All investigations provide opportunities to use the *FOSS Science Resources* as a source from which to draw evidence to support their ideas (e.g., discussion questions at the end of the articles).<br><br>Selected examples<br>Inv 1, Part 1, Steps 26, 34; Inv 1, Part 3, Step 14<br>Inv 2, Part 1, Steps 12, 23; Inv 2, Part 2, Steps 18, 20, 22<br>Inv 2, Part 3, Step 22; Inv 2, Part 4, Step 18<br>Inv 3, Part 1, Step 15; Inv 3, Part 2, Step 17; Inv 3, Part 4, Step 18 | All investigations provide opportunities to use the *FOSS Science Resources* as a source from which to draw evidence to support their ideas (e.g., discussion questions at the end of the articles).<br><br>Selected examples<br>Inv 2, Part 4, Step 19 |

# SPEAKING AND LISTENING STANDARDS

| | Standard | Energy Module |
|---|---|---|
| **Comprehension and Collaboration** | 1. Engage effectively in a range of collaborative discussions (one-on-one, in groups, and teacher-led) with diverse partners on *grade 4 topics and texts*, building on others' ideas and expressing their own clearly.<br>  a. Come to discussions prepared, having read or studied required material; explicitly draw on that preparation and other information known about the topic to explore ideas under discussion.<br>  b. Follow agreed-upon rules for discussions and carry out assigned roles.<br>  c. Pose and respond to specific questions to clarify or follow up on information, and make comments that contribute to the discussion and link to the remarks of others.<br>  d. Review the key ideas expressed and explain their own ideas and understanding in light of the discussion. | All investigations provide students ample opportunities to engage in a range of collaborative discussions.  Students discuss before, during, and after the active investigation and during the Wrap-up/Warm-up section.<br>Selected examples<br>Inv 1, Part 1, Steps 9, 13, 16, 22, 23; Inv 1, Part 2, Steps 8, 13, 17-19, 24, 27; Inv 1, Part 3, Steps 1, 6, 7, 9, 14, 16, 17, 23, 31; Inv 1, Part 4, Step 4, 5, 7, 8<br>Inv 2, Part 1, Step 27;  Inv 2, Part 2, Steps 10, 13, 15, 20, 30;  Inv 2, Part 3, Steps 3-5, 11, 13<br>Inv 3, Part 1, Steps 10, 13, 20<br>Inv 3, Part 2, Steps 2, 9; Inv 3, Part 2, Step 17<br>Inv 3, Part 3, Step 18<br>Inv 4, Part 1, Step 10; Inv 4, Part 2, Steps 5, 7<br>Inv 4, Part 3, Steps 4, 10, 15<br>Discuss articles in *FOSS Science Resources*<br>All investigations provide students with the opportunity to discuss the readings in pairs, small groups, and whole class.<br>Selected examples<br>Inv 1, Part 1, Step 21; Inv 1, Part 3, Steps 28, 29<br>Inv 1, Part 4, Step 13<br>Inv 2, Part 2, Step 26; Inv 2, Part 3, Step 20<br>Inv 3, Part 1, Step 19; Inv 3, Part 3, Steps 14, 15<br>Inv 4, Part 1, Step 15; Inv 4, Part 3, Step 19<br>Inv 5, Part 1, Step 20; Inv 5, Part 2, Steps 24, 27, 30 |
| | 2. Paraphrase portions of a text read aloud or information presented in diverse media and formats, including visually, quantitatively, and orally. | Discuss articles read aloud in *FOSS Science Resources* and online activities.<br><br>Selected examples<br>Inv 1, Part 1, Steps 15, 18<br>Inv 1, Part 2, Steps 21, 23; Inv 1, Part 4, Step 11<br>Inv 2, Part 1, Step 26; Inv 2, Part 2, Step 29<br>Inv 3, Part 1, Step 12; Inv 3, Part 2, Steps 12, 16<br>Inv 4, Part 3, Step 19<br><br>Video discussions<br>Inv 2, Part 2, Step 28<br>Inv 4, Part 3, Step 26<br>Inv 5, Part 1, Step 15; Inv 5, Part 2, Step 18 |

| Environments Module | Soils, Rocks, and Landforms Module |
|---|---|
| All investigations provide students ample opportunities to engage in a range of collaborative discussions. Students discuss before, during, and after the active investigation and during the Wrap-up/Warm-up section. | All investigations provide students ample opportunities to engage in a range of collaborative discussions. Students discuss before, during, and after the active investigation and during the Wrap-up/Warm-up section. |

**Environments Module**

Selected examples
Inv 1, Part 1, Steps 11; 34; Inv 1, Part 2, Step 6
Inv 2, Part 1, Steps 1, 11, 12, 16, 17, 25
Inv 2, Part 2, Steps 25-27; Inv 2, Part 3, Step 17
Inv 3, Part 1, Steps 4, 6, 9, 16; Inv 3, Part 2, Steps 10-12, 30-31; Inv 3, Part 3, Steps 4, 9, 16
Inv 4, Part 1, Steps 12, 17, 21, 39, 51; Inv 4, Part 2, Step 19

Discuss articles in *FOSS Science Resources*
All investigations provide students with the opportunity to discuss the readings in pairs, small groups, and whole class.

Selected examples
Inv 1, Part 1, Steps 18, 33; Inv 1, Part 3, Step 14
Inv 2, Part 1, Step 23; Inv 2, Part 2, Steps 20-23
Inv 2, Part 3, Step 22; Inv 2, Part 4, Step 20
Inv 3, Part 1, Steps 14, 15; Inv 3, Part 2, Steps 9, 17
Inv 3, Part 3, Step 13; Inv 3, Part 4, Step 18
Inv 4, Part 1, Step 50; Inv 4, Part 2, Step 17

**Soils, Rocks, and Landforms Module**

Selected examples
Inv 1, Part 1, Steps 7, 13, 17, 21; Inv 1, Part 2, Steps 3, 14, 15
Inv 1, Part 3, Steps 7, 8, 15, 16; Inv 1, Part 4, Steps 3, 14
Inv 2, Part 1, Step 7; Inv 2, Part 2, Steps 3, 16-21, 28
Inv 2, Part 3, Steps 9, 13; Part 4, Step 12
Inv 3, Part 1, Step 23; Inv 3, Part 2, Step 8
Inv 3, Part 3, Steps 1-3, 12, 14
Inv 4, Part 1, Steps 3, 15; Inv 4, Part 3, Steps 9-11

Discussing articles in *FOSS Science Resources*
All investigations provide students with the opportunity to discuss the readings in pairs, small groups, and whole class.

Selected examples
Inv 1, Part 1, Steps 19-20; Inv 1, Part 3, Steps 20, 24
Inv 2, Part 1, Steps 20, 21; Inv 2, Part 4, Steps 18, 20, 22
Inv 4, Part 2, Step 16

---

**Environments Module**

Discuss articles read aloud in *FOSS Science Resources* and online activities.

Selected examples
Inv 1, Part 3, Step 13
Inv 2, Part 3, Step 26
Inv 3, Part 2, Steps 14, 29; Inv 3, Part 3, Step 15
Inv 3, Part 4, Step 18
Inv 4, Part 1, Step 46

Video discussions
Inv 2, Part 4, Steps 13, 16
Inv 4, Part 3, Step 5

**Soils, Rocks, and Landforms Module**

Discuss articles read aloud in *FOSS Science Resources* and online activities.

Selected examples
Inv 1, Part 1, Step 19; Inv 1, Part 3, Step 20
Inv 2, Part 1, Step 19; Inv 2, Part 4, Step 18

Video discussions
Inv 1, Part 3, Step 23; Inv 1, Part 4, Step 16
Inv 2, Part 2, Steps 26, 27; Part 4, Steps 5, 6
Inv 3, Part 2, Step 16; Inv 3, Part 3, Steps 7, 11
Inv 4, Part 1, Steps 5, 14

# SPEAKING AND LISTENING STANDARDS (*continued*)

| Standard | Energy Module |
|---|---|
| **Comprehension and Collaboration** | |
| 3. Identify the reasons and evidence a speaker provides to support particular points. | All investigations provide students with the opportunity to identify the reasons and evidence their peers and the teacher provide to support their ideas (e.g. students compare their responses to the focus question and the response sheets during the Wrap-up/Warm-up section).<br><br>Selected examples<br>Inv 1, Part 1, Step 10; Inv 1, Part 3, Steps 19, 20<br>Inv 4, Part 3, Step 19<br><br>Other opportunities arise when students present information to their group or the whole class. |
| **Presentation of Knowledge and Ideas** | |
| 4. Report on a topic or text, tell a story, or recount an experience in an organized manner, using appropriate facts and relevant, descriptive details to support main ideas or themes; speak clearly at an understandable pace. | All investigations provide students with the opportunity to report on their results. In the Wrap-up/Warm-up section students recount what they did in the investigation and share their answers to the focus question. Students report on what they learn from the text when discussing the articles in *FOSS Science Resources*.<br><br>Selected examples<br>Inv 1, Part 1, Steps 1, 3, 7, 19; Inv 1, Part 2, Steps 1, 2, 6, 11, 26; Inv 1, Part 3, Steps 10, 29<br>Inv 1, Part 4, Steps 2, 8<br>Inv 2, Part 1, Steps 6, 7, 12, 16, 24; Inv 2, Part 2, Steps 3, 11, 14, 16; Inv 2, Part 3, Steps 1, 6, 16, 20<br>Inv 3, Part 1, Steps 3, 7; Inv 3, Part 2, Step 6<br>Inv 3, Part 3, Steps 6, 7, 14<br>Inv 4, Part 1, Steps 1, 5, 12; Inv 4, Part 2, Steps 4, 10<br>Inv 4, Part 3, Steps 5, 11, 17<br>Inv 5, Part 1, Step 9; Inv 5, Part 3, Steps 6, 7 |

| Environments Module | Soils, Rocks, and Landforms Module |
|---|---|
| All investigations provide students with the opportunity to identify the reasons and evidence their peers and the teacher provide to support their ideas (e.g., students compare their responses to the focus question and the response sheets during the Wrap-up/Warm-up section). | All investigations provide students with the opportunity to identify the reasons and evidence their peers and the teacher provide to support their ideas (e.g., students compare their responses to the focus question and the response sheets during the Wrap-up/Warm-up section). |

Selected examples
Inv 1, Part 1, Step 34
Inv 2, Part 3, Step 27
Inv 3, Part 2, Step 31; Inv 3, Part 3, Step 16
Inv 4, Part 2, Step 19

Other opportunities arise when students present information to their group or the whole class.

Selected examples
Inv 1, Part 2, Steps 11, 15; Inv 1, Part 3, Step 20
Inv 2, Part 2, Step 28

Other opportunities arise when students present information to their group or the whole class.

---

All investigations provide students with the opportunity to report on their results. In the Wrap-up/Warm-up section students recount what they did in the investigation and share their answers to the focus question. Students report on what they learn from the text when discussing the articles in *FOSS Science Resources*.

Selected examples
Inv 1, Part 1, Steps 1, 5, 31, 32; Inv 1, Part 2, Steps 3, 4
Inv 1, Part 2, Steps 14-16, 27, 29, 31, 3
Inv 1, Part 3, Step 9
Inv 2, Part 1, Steps 2, 4, 7, 21; Inv 2, Part 2, Step 20
Inv 2, Part 3, Steps 1, 24, 25; Inv 2, Part 4, Steps, 1, 2, 10, 18, 20
Inv 3, Part 2, Steps 9, 15, 17; Inv 3, Part 3, Steps 1, 6
Inv 3, Part 4, Steps 1, 12, 14
Inv 4, Part 1, Steps 1, 3, 15, 30, 31, 33, 45
Inv 4, Part 2, Steps 1, 11, 12; Inv 4, Part 3, Step 1

All investigations provide students with the opportunity to report on their results. In the Wrap-up/Warm-up section students recount what they did in the investigation and share their answers to the focus question. Students report on what they learn from the text when discussing the articles in *FOSS Science Resources*.

Selected examples
Inv 1, Part 1, Steps 2, 3, 20; Inv 1, Part 2, Steps 4, 7
Inv 1, Part 3, Step 19; Inv 1, Part 4, Step 1
Inv 2, Part 1, Step 12; Inv 2, Part 3, Steps 1, 2, 7-9; Part 4, Steps 2, 3, 18
Inv 3, Part 1, Step 1; Inv 3, Part 2, Steps 13-15; Inv 3, Part 3, Steps 9, 12; Inv 3, Part 4, Step 5; Inv 3, Part 4, Steps 10, 11
Inv 4, Part 1, Steps 1, 9, 12, 13; Inv 4, Part 2, Steps 10, 11
Inv 4, Part 3, Step 1; Inv 4, Part 3, Steps 12, 14, 16

## SPEAKING AND LISTENING STANDARDS (*continued*)

| Standard | Energy Module |
|---|---|
| **Presentation of Knowledge and Ideas** 5. Add audio recordings and visual displays to presentations when appropriate to enhance the development of main ideas or themes. | Inv 1, Part 2, Step 26; Inv 1, Part 3, Steps 18, 31; Inv 1, Language Extensions. Make a Poster<br>Inv 2, Part 3, Steps 16, 20<br>Inv 3, Part 1, Step 19; Inv 3, Part 2, Step 16<br>Inv 4, Part 2, Step 19 |
| 6. Differentiate between contexts that call for formal English (e.g., presenting ideas) and situations where informal discourse is appropriate (e.g., small-group discussion); use formal English when appropriate to task and situation. (See grade 4 Language standard 1 for specific expectations.) | All investigations provide students with situations in which they use either informal (small group discussions) or formal discourse structures and procedures (whole-group sharing). Protocols and sentence frames are provided for students who need support.<br><br>Selected examples<br>Inv 1, Part 2, Step 26<br>Inv 2, Part 2, Step 26<br>Inv 4, Part 2, Step 19 |

| Environments Module | Soils, Rocks, and Landforms Module |
|---|---|
| Inv 2, Part 3, Step 27; Inv 2, Part 4, Step 18<br>Inv 3, Part 2, Step 13 | Inv 2, Part 1, Step 20; Inv 2, Part 4, Step 18 |
| All investigations provide students with situations in which they use either informal (small group discussions) or formal discourse structures and procedures (whole-group sharing). Protocols and sentence frames are provided for students who need support.<br><br>Selected examples<br>Inv 1, Part 2, Step 32<br>Inv 2, Part 2, Step 20; Inv 2, Part 4, Steps 18, 20<br>Inv 3, Part 2, Steps 13, 17 | All investigations provide students with situations in which they use either informal (small group discussions) or formal discourse structures and procedures (whole-group sharing). Protocols and sentence frames are provided for students who need support.<br><br>Selected examples<br>Inv 1, Part 1, Steps 2-3, 20; Inv 1, Part 2, Steps 14, 15<br>Inv 2, Part 4, Step 18 |

# LANGUAGE STANDARDS

| Standard | Energy Module |
|---|---|
| **Conventions of Standard English**<br><br>1. Demonstrate command of the conventions of standard English grammar and usage when writing or speaking.<br>  a. Use relative pronouns (*who, whose, whom, which, that*) and relative adverbs (*where, when, why*).<br>  b. Form and use the progressive (e.g., *I was walking; I am walking; I will be walking*) verb tenses.<br>  c. Use modal auxiliaries (e.g., *can, may, must*) to convey various conditions.<br>  d. Order adjectives within sentences according to conventional patterns (e.g., *a small red bag* rather than a *red small bag*).<br>  e. Form and use prepositional phrases.<br>  f. Produce complete sentences, recognizing and correcting inappropriate fragments and run-ons.<br>  g. Correctly use frequently confused words (e.g., *to, too, two; there, their*). | All investigations provide opportunities for students to apply the conventions of English grammar when writing and speaking.<br><br>Selected examples<br>Inv 1, Part 1, Step 16; Inv 1, Part 2, Step 22<br>Inv 2, Part 3, Steps 7, 17<br>Inv 3, Part 2, Step 10<br>Inv 4, Part 2, Step 4; Inv 4, Part 3, Step 19 |
| 2. Demonstrate command of the conventions of standard English capitalization, punctuation, and spelling when writing.<br>  a. Use correct capitalization.<br>  b. Use commas and quotation marks to mark direct speech and quotations from a text.<br>  c. Use a comma before a coordinating conjunction in a compound sentence.<br>  d. Spell grade-appropriate words correctly, consulting references as needed. | All investigations provide opportunities for students to demonstrate command of the conventions of standard English capitalization, punctuation, and spelling when writing in their science notebooks, response sheets, and the I-Checks. |
| **Knowledge of Language**<br><br>3. Use knowledge of language and its conventions when writing, speaking, reading, or listening.<br>  a. Choose words and phrases to convey ideas precisely.<br>  b. Choose punctuation for effect.<br>  c. Differentiate between contexts that call for formal English (e.g., presenting ideas) and situations where informal discourse is appropriate (e.g., small-group discussion). | All investigations provide opportunities for students to use their knowledge of language and its conventions when writing in their science notebooks, discussing the investigation, and reading the articles in *FOSS Science Resources*.<br><br>Selected examples<br>Inv 1, Part 1, Steps 3, 23; Inv 1, Part 2, Step 22<br>Inv 2, Part 1, Step 27; Inv 2, Part 2, Step 26<br>Inv 2, Part 3, Step 17<br>Inv 3, Part 1, Step 19<br>Inv 4, Part 3, Step 25 |

| Environments Module | Soils, Rocks, and Landforms Module |
|---|---|
| All investigations provide opportunities for students to apply the conventions of English grammar when writing and speaking.<br><br>Selected examples<br>Inv 1, Part 2, Step 19, 32; Inv 1, Part 3, Step 10<br>Inv 2, Part 1, Step 14; Inv 2, Part 2, Steps 12, 23<br>Inv 2, Part 3, Steps 17, 19; Inv 2, Part 4, Step 14<br>Inv 3, Part 2, Steps 3, 26, 30; Inv 3, Part 4, Step 18<br>Inv 4, Part 1, Steps 43, 47, 50; Inv 4, Part 2, Step 14 | All investigations provide opportunities for students to apply the conventions of English grammar when writing and speaking.<br><br>Selected examples<br>Inv 1, Part 1, Step 16; Inv 1, Part 2, Steps 11, 13, 15<br>Inv 1, Part 3, Steps 19, 22<br>Inv 2, Part 1, Step 14; Inv 2, Part 2, Step<br>Inv 2, Part 2, Steps 7, 15, 20, 23, 24<br>Inv 2, Part 3, Step 10; Inv 2, Part 4, Steps 14, 18<br>Inv 4, Part 1, Step 7; Inv 4, Part 2, Step 14 |
| All investigations provide opportunities for students to demonstrate command of the conventions of standard English capitalization, punctuation, and spelling when writing in their science notebooks, response sheets, and the I-Checks. | All investigations provide opportunities for students to demonstrate command of the conventions of standard English capitalization, punctuation, and spelling when writing in their science notebooks, response sheets, and the I-Checks. |
| All investigations provide opportunities for students to use their knowledge of language and its conventions when writing in their science notebooks, discussing the investigation, and reading the articles in *FOSS Science Resources*.<br><br>Selected examples<br>Inv 1, Part 1, Steps 20, 21, 23; Inv 1, Part 2, Step 16<br>Inv 2, Part 2, Step 23; Inv 2, Part 3, Steps 17, 19, 20<br>Inv 3, Part 4, Step 18<br>Inv 4, Part 1, Steps 48, 50 | All investigations provide opportunities for students to use their knowledge of language and its conventions when writing in their science notebooks, discussing the investigation, and reading the articles in *FOSS Science Resources*.<br><br>Selected examples<br>Inv 1, Part 1, Steps 2-3; Inv 1, Part 2, Steps 14, 15<br>Inv 2, Part 2, Steps 6, 28; Inv 2, Part 4, Step 18 |

## LANGUAGE STANDARDS (*continued*)

| Standard | Energy Module |
|---|---|
| 4. Determine or clarify the meaning of unknown and multiple-meaning words and phrases based on *grade 4 reading and content*, choosing flexibly from a range of strategies.<br>  a. Use context (e.g., definitions, examples, or restatements in text) as a clue to the meaning of a word or phrase.<br>  b. Use common, grade-appropriate Greek and Latin affixes and roots as clues to the meaning of a word (e.g., *telegraph, photograph, autograph*).<br>  c. Consult reference materials (e.g., dictionaries, glossaries, thesauruses), both print and digital, to find the pronunciation and determine or clarify the precise meaning of key words and phrases. | All investigations provide opportunities for students to determine or clarify meaning of academic and science-specific words and phrases while reading and discussing articles in *FOSS Science Resources*.<br><br>Selected examples<br>Inv 1, Part 1, Steps 19, 21<br>Inv 4, Part 3, Steps 21, 22, 25 |
| 5. Demonstrate understanding of figurative language, word relationships, and nuances in word meanings.<br>  a. Explain the meaning of simple similes and metaphors (e.g., *as pretty as a picture*) in context.<br>  b. Recognize and explain the meaning of common idioms, adages, and proverbs.<br>  c. Demonstrate understanding of words by relating them to their opposites (antonyms) and to words with similar but not identical meanings (synonyms). | Students learn the word relationships (e.g., concept maps) and nuances of certain words that have a specific meaning in science, such as **battery, cell, circuit, component, conductor, contact, electricity, energy, source, heat, light, motion, open, closed, switch, system, terminal, transfer, attract, compass, force, gravity, interact, pole, repel, induced, permanent, code, coil, core, key, predict, rivet, collide, friction, kinetic, potential, stationary, transfer, absorb, color, frequency, mirror, peak, ray reflect, troughs, transmit, wave**.<br><br>Selected examples<br>Inv 1, Part 2, Steps 24-25<br>Inv 1, Language Extensions. Name that insulator/conductor<br>Inv 4, Part 3, Step 25<br>Inv 5, Part 2, Step 30 |

**Vocabulary Acquisition and Use**

| Environments Module | Soils, Rocks, and Landforms Module |
|---|---|
| All investigations provide opportunities for students to determine or clarify meaning of academic and science-specific words and phrases while reading and discussing articles in *FOSS Science Resources*.<br><br>Selected examples<br>Inv 1, Part 1, Steps 16-18<br>Inv 1, Part 2, Step 16<br>Inv 2, Part 1, Step 2 EL Note, Step 17<br>Inv 2, Part 2, Steps 1, 8, 18<br>Inv 2, Part 4, Steps 16-20<br>Inv 3, Part 1, Steps 14, 15, 25<br>Inv 4, Part 2, Step 17 | All investigations provide opportunities for students to determine or clarify meaning of academic and science-specific words and phrases while reading and discussing articles in *FOSS Science Resources*.<br><br>Selected examples<br>Inv 1, Part 1, Step 19<br>Inv 2, Part 1, Step 8<br>Inv 4, Part 1, Step 11<br>Inv 4, Part 2, Step 11 |
| Students learn the word relationships and nuances of certain words that have a specific meaning in science, such as **adult, behavior, condition, environment, factor, function, inference, living, nonliving, observation, organism, preferred, stage, structure, carrying capacity, competition, consumer, decomposer, ecosystem, energy, food chain, food web, home range, interaction, population, producer, mimicry, brine, concentration, inherited, trait, optimum, range, tolerance, reproduce, thrive, variation, adaptation, dominant, drought, distribution**.<br><br>Selected examples<br>Inv 1, Part 1, Step 27<br>Inv 2, Part 1, Steps 17, 21, 22<br>Inv 2, Part 2, Steps 7-10, 14, 18;<br>Inv 2, Part 4, Step 18<br>Inv 4, Part 2, Step 17 | Students learn the word relationships (e.g., concept maps) and nuances of certain words that have a specific meaning in science, such as **abrasion, acid, chemical reaction, weathering, conglomerate, expand, marble, system, soil, clay, humus, silt, sand, pebble, rock, basin, cast, delta, deposition, erosion, flood, fossil, imprint, meander, mold, slope, superposition, valley, contour, crust, elevation, lava, magma, mantle, profile, aggregate, cement, concrete, fuel, power, renewable, nonrenewable, resource, solar**.<br><br>Selected examples<br>Inv 1, Part 1, Step 19<br>Inv 1, Part 2, Step 8<br>Inv 1, Part 3, Step 16<br>Inv 1, Part 3, Step 21<br>Inv 2, Part 2, Step 1<br>Inv 4, Part 1, Step 2<br>Inv 4, Part 2, Step 13 |

## LANGUAGE STANDARDS (*continued*)

| Standard | Energy Module |
|---|---|
| **Vocabulary Acquisition and Use** 6. Acquire and use accurately grade-appropriate general academic and domain-specific words and phrases, including those that signal precise actions, emotions, or states of being (e.g., *quizzed, whined, stammered*) and that are basic to a particular topic (e.g., *wildlife, conservation,* and *endangered* when discussing animal preservation). | All investigations provide opportunities for students to acquire and use conversational, academic, and science-specific words and phrases. Science vocabulary words are in bold when they are first introduced to students in *FOSS Science Resources.* Students also review the vocabulary in the Review vocabulary section for each part and the Wrap-up review vocabulary section of each investigation. Selected examples Inv 1, Parts 1-4, 8, 10, 11, 14, 16, 22 Inv 1, Part 2, Steps 1, 3, 6, 11, 12, 19, 21, 22, 27 Inv 1, Part 3, Steps 5, 12, 14, 19, 23, 25-27, 31 Inv 2, Part 1, Steps 3, 12, 13, 16, 17 Inv 2, Part 2, Steps 1, 3, 4, 6, 8, 12, 16, 17, 21-23 Inv 2, Part 3, Steps 1, 10 Inv 3, Part 1, Steps 4, 8, 10, 14, 15 Inv 3, Part 3, Steps 2, 10 Inv 4, Part 1, Steps 5, 16 Inv 4, Part 2, Steps 11, 13, 14, 19 Inv 4, Part 3, Steps 1, 2, 6, 13, 14 Inv 5, Part 1, Steps 1, 5, 10, 11, 16 Inv 5, Part 2, Steps 9, 15, 19 Inv 5, Part 3, Steps 2, 8 |

| **Environments Module** | **Soils, Rocks, and Landforms Module** |
|---|---|
| All investigations provide opportunities for students to acquire and use conversational, academic, and science-specific words and phrases. Science vocabulary words are in bold when they are first introduced to students in *FOSS Science Resources*. Students also review the vocabulary in the Review vocabulary section for each part and the Wrap-up Review vocabulary section of each investigation. | All investigations provide opportunities for students to acquire and use conversational, academic, and science-specific words and phrases. Science vocabulary words are in bold when they are first introduced to students in *FOSS Science Resources*. Students also review the vocabulary in the Review vocabulary section for each part and the Wrap-up Review vocabulary section of each investigation. |
| Selected examples<br>Inv 1, Part 1, Steps 1, 9, 20, 21, 23, 24, 26, 27<br>Inv 1, Part 2, Steps 1, 3, 5, 14, 18, 32, 35-37<br>Inv 2, Part 1, Steps 2, 8, 15, 16-19, 23-25<br>Inv 2, Part 2, Steps 1, 3, 6-1<br>Inv 2, Part 3, Steps 2, 17-19<br>Inv 2, Part 4, Steps 18-20<br>Inv 3, Part 1, Steps 1, 2, 6, 12, 13<br>Inv 3, Part 3, Steps 2, 7<br>Inv 3, Part 4, Steps 1, 14-16<br>Inv 4, Part 1, Steps 39-43<br>Inv 4, Part 2, Steps 1, 11, 13, 14<br>Inv 4, Part 3, Step 1 | Selected examples<br>Inv 1, Part 1, Steps 1, 8, 15, 16<br>Inv 1, Part 2, Steps 12-13<br>Inv 1, Part 3, Steps 3, 16, 21, 22<br>Inv 2, Part 1, Steps 1, 8-10, 13, 14, 21<br>Inv 2, Part 2, Steps 1, 6, 7, 14, 15, 28<br>Inv 2, Part 3, Step 1<br>Inv 2, Part 4, Steps 1-3, 5,6, 12-14<br>Inv 3, Part 1, Steps 2, 5, 8, 12, 13, 17, 18<br>Inv 3, Part 2, Steps 1, 9, 10, 17, 19<br>Inv 3, Part 4, Step 5<br>Inv 4, Part 1, Steps 3, 5, 7<br>Inv 4, Part 2, Steps 2, 4, 13, 14, 16 |

# FOSS and Common Core Math – Grade 4

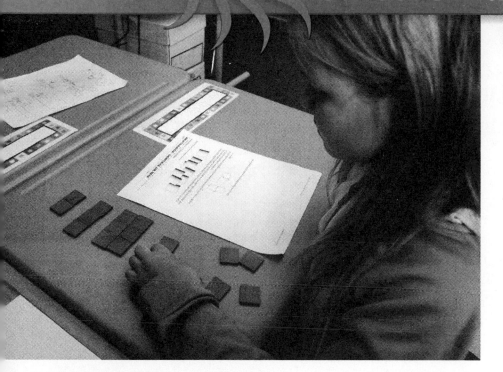

## Contents

Introduction ..............................1

Operations and Algebraic Thinking ...................................4

Number and Operations in Base 10 .....................................6

Number and Operations in Fractions ...................................8

Measurement and Data ..........10

# INTRODUCTION

The adoption of the Common Core State Standards for Mathematics calls for shifts in focus, coherence, and rigor. The teaching of the standards should be focused on the important content, coherent from one grade level to the next, and rigorous in requiring conceptual understanding, fluency, and application. Within this area of application, FOSS provides fertile ground for the use of mathematics.

The FOSS Program integrates mathematics with science in two ways throughout the grade 4 modules. In active investigations, students apply mathematics during data gathering and analysis. In addition, the Interdisciplinary Extensions at the end of each investigation usually include a math problem of the week. These problems enhance the science learning by providing hypothetical data for students to analyze or in some way relate to the context of the investigation. The notes explain for the teacher the problem and describe how students might approach its solution. The problems are prepared for distribution to students on duplication masters in the Teacher Masters chapter of *Teacher Resources*.

This chapter gives an overview of how FOSS addresses the Common Core State Standards for Mathematics through science. It also points out specific instances in which students exercise those skills during science instruction.

## Mathematical Practices

Mathematical practices consist of eight processes and proficiencies that are important for all students.

1. Make sense of problems and persevere in solving them.

2. Reason abstractly and quantitatively.

3. Construct viable arguments and critique the reasoning of others.

4. Model with mathematics.

5. Use appropriate tools strategically.

6. Attend to precision.

7. Look for and make use of structure.

8. Look for and express regularity in repeated reasoning.

Within the context of science, students use some of these mathematical practices on a regular basis. According to *Next Generation Science Standards* (volume 2, appendix L, p. 138),

The three CCSSM practice standards most directly relevant to science are:

- MP.2. Reason abstractly and quantitatively.

- MP.4. Model with mathematics.

- MP.5. Use appropriate tools strategically.

When students reason abstractly and quantitatively and model with mathematics, they are using math in context. They work with symbols and their meanings and represent and solve word problems. Students choose and correctly use the available tools to collect data and solve problems. In the grade 4 modules, students engage with these three practices during the active investigation and by completing the problems at the end of each investigation. Here are some examples.

In solving the Math Problem of the week for Investigation 3 of the **Environments Module**, students model a particular recipe using a given amount of water and changing amount of salt. In order to solve this, students reason quantitatively and use mathematics in the context of science. They can use drawings as they multiply or divide fractional amounts of salt in different recipes.

In the **Soils, Rocks, and Landforms Module**, students are presented with a challenge about a traditional game, Rock, Paper, Scissors. The problem appears simple, but requires students to create a model to determine how many games were played. Story problems provide opportunities for students to utilize tools to determine the solutions to two-step problems.

In the **Energy Module**, students are asked to reason abstractly to determine the number of wires needed to create series or parallel circuits for a given number of bulbs. This requires students to determine the type of pattern that exists between the number of bulbs and wires for each type of circuit and represent that pattern mathematically. Using this pattern, they can then determine the number of wires or bulbs needed.

## Mathematical Content

The mathematical content in fourth grade is organized around four concepts.

- Operations and algebraic thinking
- Number and operations in base 10
- Number and operations in fractions
- Measurement and data

The following pages have a table that identifies the opportunities to engage students in developing these mathematical concepts as well as those learned in grade 3. It lists the math content for some third and fourth grades and points out relevant opportunities in the three FOSS modules for grade 4.

# OPERATIONS AND ALGEBRAIC THINKING

| Standard | Energy Module |
|---|---|
| **Use the four operations with whole numbers to solve problems.** | |
| 3. Solve multistep word problems posed with whole numbers and having whole-number answers using the four operations, including problems in which remainders must be interpreted. Represent these problems using equations with a letter standing for the unknown quantity. Assess the reasonableness of answers using mental computation and estimation strategies including rounding. | |
| **Generate and analyze patterns.** | |
| 5. Generate a number or shape pattern that follows a given rule. Identify apparent features of the pattern that were not explicit in the rule itself. *For example, given the rule "Add 3" and the starting number 1, generate terms in the resulting sequence and observe that the terms appear to alternate between odd and even numbers. Explain informally why the numbers will continue to alternate in this way.* | Inv 1, Problem of the week |

Grade 4

From *Common Core State Standards for English Language Arts and Literacy in History/Social Studies, Science, and Technical Subjects* (by National Governors Association Center for Best Practices and Council of Chief State School Officers, (Washington, DC: authors, 2010).

| Environments Module | Soils, Rocks, and Landforms Module |
|---|---|
| Inv 2, Problem of the week | Inv 4, Problem of the week |
| Inv 4, Problem of the week | |

## NUMBER AND OPERATIONS IN BASE 10

| Standard | Energy Module |
|---|---|
| **Use place value understanding and properties of operations to perform multi-digit arithmetic.** | |
| 5. Multiply a whole number of up to four digits by a one-digit whole number, and multiply two two-digit numbers, using strategies based on place value and the properties of operations. Illustrate and explain the calculation by using equations, rectangular arrays, and/or area models. | |

Grade 4

| Environments Module | Soils, Rocks, and Landforms Module |
|---|---|
| | |
| Inv 1, Problem of the week | |

# NUMBER AND OPERATIONS IN FRACTIONS

| Standard | Energy Module |
|---|---|
| **Build fractions from unit fractions by applying and extending previous understandings of operations on whole numbers.** | |
| 3. Understand a fraction $a/b$ with $a > 1$ as a sum of fractions $1/b$.<br>  a. Understand addition and subtraction of fractions as joining and separating parts referring to the same whole.<br>  b. Decompose a fraction into a sum of fractions with the same denominator in more than one way, recording each decomposition by an equation. Justify decompositions, e.g., by using a visual fraction model. *Examples: 3/8 = 1/8 + 1/8 + 1/8; 3/8 = 1/8 + 2/8 ; 2 1/8 = 1 + 1 + 1/8 = 8/8 + 8/8 + 1/8.*<br>  c. Add and subtract mixed numbers with like denominators, e.g., by replacing each mixed number with an equivalent fraction, and/or by using properties of operations and the relationship between addition and subtraction.<br>  d. Solve word problems involving addition and subtraction of fractions referring to the same whole and having like denominators, e.g., by using visual fraction models and equations to represent the problem. | |
| 4. Apply and extend previous understandings of multiplication to multiply a fraction by a whole number.<br>  a. Understand a fraction $a/b$ as a multiple of $1/b$. *For example, use a visual fraction model to represent 5/4 as the product 5 × (1/4), recording the conclusion by the equation 5/4 = 5 × (1/4).*<br>  b. Understand a multiple of $a/b$ as a multiple of $1/b$, and use this understanding to multiply a fraction by a whole number.<br>  c. Solve word problems involving multiplication of a fraction by a whole number, e.g., by using visual fraction models and equations to represent the problem. | |

Grade 4

| Environments Module | Soils, Rocks, and Landforms Module |
|---|---|
| | |
| Inv 3, Problem of the week | |
| Inv 3, Problem of the week | |

## MEASUREMENT AND DATA

| Standard | Energy Module |
|---|---|
| **Solve problems involving measurement and estimation of intervals of time, liquid volumes, and masses of objects.** | |
| 2. Measure and estimate liquid volumes and masses of objects using standard units of grams (g), kilograms (kg), and liters (l). Add, subtract, multiply, or divide to solve one-step word problems involving masses or volumes that are given in the same units, e.g., by using drawings (such as a beaker with a measurement scale) to represent the problem. | |
| **Represent and interpret data.** | |
| 3. Draw a scaled picture graph and a scaled bar graph to represent a data set with several categories. Solve one- and two-step "how many more" and "how many less" problems using information presented in scaled bar graphs. *For example, draw a bar graph in which each square in the bar graph might represent 5 pets.* | |
| **Geometric measurement: recognize perimeter as an attribute of plane figures and distinguish between linear and area measures.** | |
| 8. Solve real world and mathematical problems involving perimeters of polygons, including finding the perimeter given the side lengths, finding an unknown side length, and exhibiting rectangles with the same perimeter and different areas or with the same area and different perimeters. | Inv 5, Problem of the week |

*(left margin: Grade 3)*

| Environments Module | Soils, Rocks, and Landforms Module |
|---|---|
| | |
| Inv 3, Part 1, Step 8, Set up the hatching experiment<br>Inv 3, Part 3, Step 4, Formulate a prediction and test it<br>Inv 4, Part 1, Step 10, Set up the water experiment<br>Inv 4, Part 1, Step 25, Water the seeds | Inv 1, Part 1, Step 9, Add water<br>Inv 1, Part 3, Step 6, Conduct the investigation<br>Inv 1, Part 4, Step 11, Discuss further separation |
| | |
| | Inv 3, Problem of the week |
| | |
| Inv 1, Problem of the week | |

## MEASUREMENT AND DATA (*continued*)

| Standard | Energy Module |
|---|---|
| **Solve problems involving measurement and conversion of measurements from a larger unit to a smaller unit.** | |
| 1. Know relative sizes of measurement units within one system of units including km, m, cm; kg, g; lb, oz.; l, ml; hr, min, sec. Within a single system of measurement, express measurements in a larger unit in terms of a smaller unit. Record measurement equivalents in a two-column table. *For example, know that 1 ft is 12 times as long as 1 in. Express the length of a 4 ft snake as 48 in. Generate a conversion table for feet and inches listing the number pairs (1, 12), (2, 24), (3, 36), ...* | |
| 2. Use the four operations to solve word problems involving distances, intervals of time, liquid volumes, masses of objects, and money, including problems involving simple fractions or decimals, and problems that require expressing measurements given in a larger unit in terms of a smaller unit. Represent measurement quantities using diagrams such as number line diagrams that feature a measurement scale. | Inv 3, Problem of the week |
| 3. Apply the area and perimeter formulas for rectangles in real world and mathematical problems. *For example, find the width of a rectangular room given the area of the flooring and the length, by viewing the area formula as a multiplication equation with an unknown factor.* | |

| Environments Module | Soils, Rocks, and Landforms Module |
|---|---|
| Inv 4, Home/School Connection | |
| | Inv 4, Problem of the week |
| Inv 1, Problem of the week<br>Inv 4, Home/School Connection | |

# Taking FOSS Outdoors

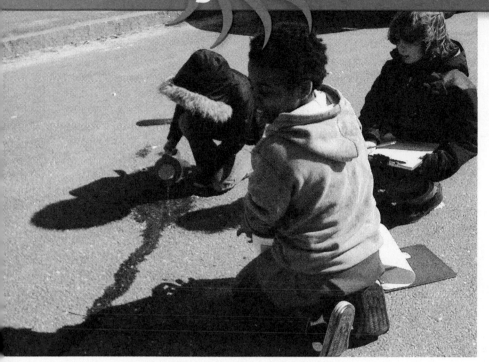

## Contents

Introduction ............................1

What Does FOSS Look Like
Outdoors? ...............................2

Goals and Objectives...............3

Managing Space .....................4

Managing Time.......................8

Managing Materials ..............10

Managing Students ...............13

Teaching Strategies ...............20

Flow of Outdoor Activities .....22

Extending beyond FOSS
Outdoor Activities ................23

Elementary-Level
Environmental Education........25

References ..........................27

Acknowledgments.................28

*If we want children to flourish, to become truly empowered, then let us allow them to love the earth before we ask them to save it.*

David Sobel, *Beyond Ecophobia*

# INTRODUCTION

During its first 20 years, FOSS focused on classroom science. The goal was to develop a scientifically literate population with an ever-growing knowledge of the natural world and the interactions and organizational models that govern and explain it. In recent years, it has become clear that we have a larger responsibility to the students we touch with our program. We have to extend classroom learning into the field to bring the science concepts and principles to life. In the process of validating classroom learning among the schoolyard trees and shrubs, down in the weeds on the asphalt, and in the sky overhead, students will develop a relationship with nature. It is our relationship with natural systems that allows us to care deeply for these systems. In order for students in our schools today to save Earth, and save it they must, they first have to feel the pulse, smell the breath, and hear the music of nature. So pack up your explorer's kit, throw open the door, and join us. We're taking FOSS outdoors.

## WHAT DOES FOSS LOOK LIKE OUTDOORS?

Visualize taking FOSS outdoors: Students exit the classroom in an orderly fashion, their direction and purpose undeterred by the joyful sounds of other students at recess. With focused enthusiasm, the band of young scientists moves toward the edge of the schoolyard. Each student is carrying something, maybe a clipboard for recording, a container for collecting, or a hand lens for observing. Students reach their destination and quickly form a sharing circle. After a brief orientation, students disperse and begin searching the tall grass along the chain-link fence. All are independently recording in their science notebooks, and all are on task. The teacher moves about with intention, speaking to a few students at a time. After several more minutes of this work, the teacher rings a chime. Students freeze, raise one arm, and look at her. She rings the chime again. Students leave their materials in their spots and re-form their sharing circle with their teacher for discussion or additional instructions.

This scenario could be anywhere in the country with a regular classroom teacher using any of the FOSS modules. Taking FOSS outdoors is a natural extension of the classroom work. It looks and feels a lot like standard FOSS activities. Many of the routines you use inside the classroom can be implemented outdoors as well. Success, however, does depend on a few specialized skills and specific preparation to maximize outdoor teaching efficiency.

Expect the enthusiasm, participation, engagement, group discussions, and effort on notebook entries to be heightened during and after an outdoor experience. Even the simplest outdoor activities create a surge of positive energy. It is difficult to determine whether the enthusiasm and commitment students exhibit when doing FOSS outdoors comes from exercising content they already know or from just being outdoors. Students can be a bit louder and more excited when they are learning outdoors, and the space allows for this expansion in energy level, which benefits some students immensely.

# GOALS AND OBJECTIVES

The three program goals set down 20 years ago still serve FOSS well. They are: (1) scientific literacy for all students, (2) instructional efficiency and support for teachers, and (3) systemic reform.

The march into the schoolyard has three objectives that relate to the goal for students. First, the outdoor activities **continue and extend the learning** that starts in the classroom. The outdoor activities provide more experience with the content and additional opportunities to practice skills and techniques developed in the classroom.

Second, venturing out provides opportunities for students to discover **applications and examples of classroom content and concepts**. The classroom activities work well for developing sound conceptual science knowledge. That knowledge, however, is constrained by the context in which the concepts are taught. For students to take the next level of ownership of that knowledge, they need to see how it applies and generalizes in the broader context of the world. Leaving the classroom context with a head full of new ideas and new tools for observation enriches the learning.

The third objective is to **connect students with nature**. On the boundaries of the planned, structured experiences are the intangibles that may spark a new relationship with natural systems. It may start with a multisensory experience in the native environment—wind, cold, sunshine, plants, insects, on and on—and advance to an awareness of the diversity of resources surrounding the school. It might evolve into a consciousness of place, followed by a flood of questions about the structure, organization, and operation of the schoolyard ecosystem. When students bond with nature, they have accepted a precious gift, and we have accomplished something important.

# MANAGING SPACE

FOSS outdoor activities are designed to be successful in a diversity of schoolyards. Some schoolyards are covered in asphalt, while others have been turned into well-designed outdoor learning environments. Some include large, grassy areas without trees, and others are covered with mulch. One outdoor space may be circled by a variety of mature trees; the next may have recently planted maples and pines scattered about. The space may reflect thoughtful attention or neglect. Nevertheless, FOSS believes that bringing students into the fresh air under a changing sky, into the available outdoor space, will awaken their well-being and stimulate their understanding of science concepts.

## Choosing Outdoor Spaces

Whether your school's landscape is wild, manicured, or asphalt, there are more options for outdoor learning spaces than might initially meet the eye. This section will help you choose the best spaces near your school for the FOSS outdoor activity.

Before choosing your outdoor study areas, get to know your outdoor spaces. Look closely at all areas surrounding the school building—even places that students do not normally go. Many seemingly uninteresting

monoculture fields are flourishing with a diversity of different grass species and other small flowering plants. Consider the pile of leaves that blew into a corner of your schoolyard; a crack in the concrete; or the ragged, weedy edge of the field where the lawn mower doesn't reach. These are places that provide small animals with what they need—food, shelter, water, and space. Transition zones where vegetation changes from shrubbery to lawn or garden to field can present interesting study sites. As you ponder the learning possibilities in and around your schoolyard, consider these characteristics.

***Accessibility.*** You should be able to walk from your classroom to your outdoor site in 2–5 minutes. Sites farther than 10 minutes away can be considered for special outings, but are not realistic for frequent access. Check out physical access if you have students whose mobility requires consideration. Be aware of slippery surfaces from water or ice, and caution students to be careful.

**Purpose.** Determine the space needs of the activity. Some activities will require open space, such as a field or blacktop. Other activities work better if students have a more diverse landscape with varying environmental conditions (such as temperature, light intensity, wind). Some activities require a variety of human-made materials to measure or test for certain properties (such as magnets to test for magnetism). Different areas will serve different needs.

**Size.** The space should be large enough for the class to work comfortably but small enough for you to supervise all students easily. You always need to be able to see all your students, and your students need to be close enough to hear you and your attention signal.

**Boundaries.** For any space you intend to use, make sure you have clearly defined the boundaries before heading outdoors with your students. Ideally, the landscape will be helpful. For example, stay between the sidewalk and the tree line. If natural markers are not present, you may need to bring along traffic cones or their equivalent to define limits. In general, consider if there are any hazards, such as dangerous debris, poisonous plants, or traffic.

## Fostering and Maintaining Diversity

For life science and earth science studies, ideally you want your site to have a variety of living and dead plant matter and a range of environmental conditions. Survey your site to see if it includes places that have been left unmanaged. Even a small wild zone along a fence or behind the maintenance area or an adjacent field can be a valuable resource. It is important for students to see that living things carry on, even in the city, without human assistance.

**Enhancing your schoolyard.** You may be able to secure a small section of the schoolyard from the school custodian, allowing it to grow wild to compare to the managed school grounds. Consult with your custodian and principal to see if this can be arranged.

Another way to enhance biodiversity is to encourage decomposition by letting fallen leaves and/or lawn clippings to remain on an area of soil over the winter. This gives worms and other decomposers something to eat, which, in turn, provides food for everything else. Make sure all necessary parties are aware of your intentions to leave an area untouched. If you find that you need administrative permission, consider ways to contain and mark the unkempt (but not unloved) area so that it clearly represents an intentional project.

***Tread lightly.*** Your schoolyard study areas will potentially experience some user impact. It is important to teach students to minimize their environmental footprint. Otherwise, the living things they disturb might seek a safer place to live. Unless the class is intentionally collecting specimens, nothing natural should be picked up or removed from the area. This is a good opportunity to introduce the "leave no trace" philosophy, which, in an effort to preserve an area for recreation, encourages us to leave natural objects as we find them.

At some schools, the outdoor space is used by so many classrooms that a system is needed to schedule outdoor activities. A sign-up sheet can be used to reserve outdoor spaces just as is done to reserve other school resources. Check the site the morning before taking the class outdoors to make sure the area is ready for students to investigate.

## Weather

Weather can present great challenges and exceptional experiences. Inclement weather can provide an excellent opportunity to study environmental concepts: water drainage, wind impact, plant and animal survival adaptations. (There is nothing like being out in a snowstorm to appreciate the value of insulation!) Making extra preparations to study out in the elements has value. If the activity can be undertaken with some assurance of success, try to make it work. Over time, students acclimate to all sorts of weather and will actually look forward to the challenge of going out in difficult weather.

***Clothing.*** The right gear at the right time can make all the difference. Baseball caps stored at school can work well in a light rain and are often essential as sun protection in warmer climates. Baseball caps in a light rain are especially helpful for students who wear glasses. If possible, invest in a set of rain ponchos to make it possible to go out in wet weather. Large trash bags can make very effective, low-cost ponchos. Of course, you will want to model this elegant attire. Communicate regularly with students and family members about upcoming outdoor experiences.

***Wind.*** A stiff breeze can fling your materials into disarray or send notebooks flying. If you anticipate wind, discuss ways to keep materials from blowing away (such as using natural paperweights or taping down nonliving specimens). If there is a protected area where you and your class can take shelter briefly, the activity can continue. You may have to chase down a couple of notebook sheets before students become accustomed to securing papers and other light materials.

## Safety and Comfort

Be prepared for the unexpected. Insect stings (ants, bees, wasps, mosquitoes) can be alarmingly painful for young students, particularly if they have not been stung before. Although extremely unlikely in a schoolyard, have a plan developed with students in advance as to how to retreat with purpose if someone accidentally disturbs a nest. You should already know who is allergic and who has never been stung before.

Skin-irritating plants (poison oak, poison ivy, poison sumac, nettles) can certainly put a damper on a field trip. Take a moment, and get to know your local irritants and toxic plants. The rule "leaves of three, let it be" works only for poison ivy and poison oak. Poison sumac has 7–13 leaves on a branch. Stinging nettle feels much like being stung by a jellyfish and can be very frightening for students who have never experienced it. Often, the irritation subsides within a few minutes; do not treat rashes with bleach or rubbing alcohol.

Lyme disease is a treatable bacterial infection, carried by deer ticks. It is present throughout the country, but is particularly present in eastern states. It is possible to get sick without finding a tick bite. If you or your students experience flulike symptoms that are severe enough to see a doctor, make sure that doctor is aware of any outdoor exposure.

If you are out and about in tick country, tuck pant legs into socks and take a few minutes at the end of the trip to pair up and look for obvious ticks on clothing and on the neck and shoulders of a partner.

# MANAGING TIME

## When to Teach

When you start a new module, anticipate when you might want to go outdoors, and schedule the time. The At a Glance chart in each investigation can help with this planning.

*Time of year.* If possible, plan the time of year when you will teach particular modules. In the northern tier, life science and earth science modules would be best in the fall or spring. In the southern tier, it might be best to teach life science modules in the winter when it is not uncomfortably hot during the day. Good times to coordinate your outdoor activities with the school calendar include minimum days or other disruptions to the regular schedule, days just before or after school vacations, and days following district testing.

*Time of day.* Consider the time of day you teach your activities. Established schedules are often difficult to alter, but you might find it advantageous to do so. If you do a lot of seat work in the mornings, you may want to break the routine occasionally with an outdoor activity. Students will return to the classroom refreshed and ready to focus on the next seated activity you have planned.

If you live in a climate where it gets really hot during the school day, you might want to teach outdoors early in the day. Conversely, if you live in a cold climate, you might want to do your winter outdoor activities midday. If you're looking for wildlife (birds, insects, mammals), the best time to go outdoors might be in the morning.

If you plan to use a part of the schoolyard that is heavily populated at predictable times during the day (lunch, physical education), plan to venture out at a time when other activities are minimal.

Stay flexible. If you are studying the **Water Module**, for example, be prepared to dash out if it rains or snows. One of the delights of outdoor education is going out when nature is putting on a show. Inquiring minds rush out for the experience when timid observers retreat.

*Specific times.* Some activities require a sunny day. Measuring shadows, solar water heaters, and solar cell investigations require sunshine. It can be tricky to move on without completing specific observations or experiments. Be creative. You may need to proceed with the module and return to the sunny-day activity when the Sun finally comes out.

## Instructional Time

An outdoor activity might require 15 minutes, or it might require an hour. Only part of the time budgeted for outdoor learning is actually spent interacting with the schoolyard terrain, plants, and animals. The rest is management.

***Travel time.*** It will take perhaps 10 minutes from the announcement that it is time to decamp for the schoolyard and the time you arrive there. It will take several minutes to describe and distribute materials, get the appropriate clothing, line up, and travel in an orderly fashion to the designated location. Travel back to the classroom will take another 3–4 minutes.

***Instructions.*** Outdoors, students form a sharing circle. It will take 2–4 minutes to review rules, set the boundaries for the activity, describe the challenge, and distribute materials.

***Investigation time.*** Students break into pairs or groups to engage in the outdoor investigation. This might be as short as 8–10 minutes or as long as 30–40 minutes.

***Wrap-up.*** Students return to the sharing circle to share and discuss their discoveries for several minutes.

***Classroom follow-up.*** Frequently, students bring artifacts back to class to display in a classroom museum or to set up for further observation.

**▶ SAFETY NOTE**
Students should not disturb or collect live organisms in their natural habitats.

Some outdoor activities call for more flexible allocations of time. An activity may call for setup early in the day with periodic monitoring or measuring throughout the day.

# MANAGING MATERIALS

When students step onto the schoolyard, they are field scientists. In the field, there is no lab bench where investigations can be set up, and there is no ready supply of materials. The field equipment must be minimal, portable, and durable so that it can be easily and safely transported from the classroom and back.

## Field Equipment

A student's outdoor bag will contain the specific materials needed for the activity of the day as well as some core necessities, such as a hand lens and a writing tool.

Student outdoor bags might contain these basics.

- Pencils/pens
- Hand lens on brightly colored string or yarn
- Colored pencils or crayons
- Measuring tape
- Vials with caps
- Clipboard or notebook
- Seat pad

Note that pens and pencils each have drawbacks: pencil points break, and pen ink freezes in extremely cold weather. Seat pads can simply be several sheets of newspaper covered with a plastic bag.

Your basic teacher's outdoor equipment bag will include a few backup student materials and some items for helping with management.

- Extra pencils, pens, hand lenses, vials, and cups
- Attention signal (chime, whistle, or cowbell)
- Tissues and paper towels
- Basic first-aid kit (adhesive bandages)
- Phone (if leaving the school grounds)
- Student class list (particularly if you teach more than one class) with appropriate student health information and/or permission slips if away from school.

## Transporting Materials

Getting materials to and from the outdoor site is a shared responsibility. Students will carry their personal equipment, and class materials can be distributed among students or tackled as a teacher task. Students always carry something to the outdoor site, even when it would be easier for you to carry everything. This reminds students that they are heading out for science, not recess. A hand lens serves as such a token.

Some teachers prefer to have students carry only their clipboards or notebooks and pencils, while the teachers carry all the field equipment in a canvas shopping bag or milk crate to the outdoor home base. Other teachers use a wagon or wheelie crate to transport the equipment. After teaching a few outdoor activities, you will discover what works best for you. Students will get excited when they see you preparing your transport system for an outdoor activity.

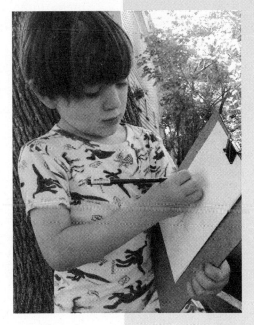

***Water.*** Water is often used during outdoor activities. If you are lucky, there will be a tap near your study site. More likely, you will carry water from the school building. Recycled plastic jugs with screw caps and smaller bottles with screw caps are good vessels.

At times, you will want open containers of water, such as buckets or basins (for washing rocks, cleaning containers, and so on). Half-filled buckets can be carried a short distance, but basins should be carried empty and filled from jugs.

You rarely have to bring water back inside. Recycle leftover water by watering schoolyard plants. Make this practice overt to help students develop respect for this vital natural resource.

## Creating Outdoor Tools

A sturdy writing surface is essential for science in the schoolyard. A bound notebook (composition book) is excellent. A serviceable clipboard can be made from a piece of cardboard and a binder clip. Use a paper cutter to cut sturdy cardboard slightly larger than a sheet of notebook paper. Place a medium-size binder clip at the top and a large elastic band around the bottom (to keep the paper from flapping up). Tie a pencil on a brightly colored string to the binder clip.

A group writing surface is important sometimes. You can use blue painter's tape to attach a sheet of chart paper temporarily to a wall or clip it onto a chain-link fence with binder clips or clothespins. On windy days, attach all four corners.

A small pack can serve as a hands-free means for students to tote their equipment. Little backpacks are excellent, but a serviceable low-cost satchel can be crafted from a large plastic bag and string. Purchase enough gallon-size zip bags for your class. Punch two holes just *under* the ends of the zipper. (This reduces tearing.) Cut the strings about 1 meter long. Tie sturdy knots that will not come undone. Store the string inside the bag after use to prevent tangling with other bags.

Hand lenses may disappear when students place them on the ground to perform a task. Run bright-colored string or yarn through holes in the lenses for students to wear. If your hand lenses do not already have holes, see if you can get holes drilled through the handles.

# MANAGING STUDENTS

Going outdoors regularly is the best way to develop a productive and joyful working relationship with students in the outdoors. When students realize that going outdoors to learn is not a special event but, rather, a science event that will occur routinely, you may be surprised at how quickly they adapt to their expanded, enriched classroom.

## Before the First Outing

It is always important to let the school administration know that you and your students will periodically be out of the classroom. If you are planning to leave the school grounds, remember to file a flight plan describing your itinerary, anticipated time of return, and contact partners.

At the beginning of the school year, send a letter home to families, letting them know that learning will extend to the schoolyard and, possibly, beyond. It may be possible to have a signed permission slip for impromptu walking field trips outside the schoolyard. Have families put their contact information and specific student health information on the permission slip. Photocopy these, and have one set of copies in the office and another set in a zip bag in your outdoor equipment bag for emergencies.

Tell students at the beginning of the year that they will be going outdoors often during science class. Remember to let them know a day in advance that they will be going outdoors. Let them know what it means to dress appropriately. This is especially important in the cold or stormy season when students will need proper clothing for safety and comfort. Your class can go out in any weather if students are dressed appropriately. A consistent system of reminders and clothing preparation will train students to be prepared.

***Safety rules.*** Creating consistent, considerate rules of engagement is important. Learning is enhanced, and behavior problems are largely averted by routines that students participate in and understand.

Have a discussion about what students think constitutes proper preparation and behavior for leaving the classroom to study outdoors. (This discussion may be most productive *after* an initial orientation

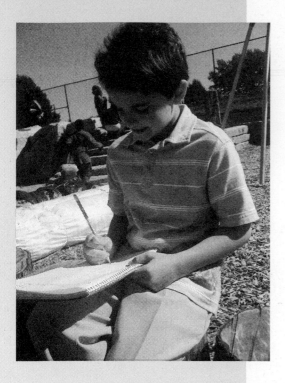

excursion to survey the schoolyard resources.) Have students generate a list of behaviors that they can adopt and respect. You may want to generate a second list of behaviors that you will agree to as leader of the adventure. Introduce as much formality into the process as you deem important. Develop the idea of a contract that all members of the class sign. Post the contract in your classroom and reference the *Outdoor Safety* poster. Here are the behaviors and rules that should appear on the list.

- Walk quickly and quietly outdoors.
- Outdoor science is not recess.
- Listen to the teacher's instructions.
- Freeze when the teacher rings the bell.
- Stay inside the boundaries.
- Don't make noise near the classrooms.
- Don't injure plants and animals in any way.
- Leave the outdoor environment the way you found it. Never release living organisms into the local environment unless they were collected there.

## First Outing

Your first trip to the schoolyard may be a bit chaotic. Students may be distracted by other activities going on, and they may lapse into recess mode. A few precautions will minimize disruptive behaviors.

***The path of egress.*** Determine which doors are available to access your outdoor site. Make sure you follow your school's policy for using auxiliary doors during the school day. Do they need to be closed at all times, or can you prop one open? Are they locked from the outside? You may be able to access keys in order to reduce travel time. When possible, avoid using the door you would normally use for recess. Students have a different mind-set when they walk onto the schoolyard through the "outdoor classroom" door.

***Orientation activity.*** Consider an orientation activity for your first outing. The stated goal might be a site tour to inventory the resources on hand. Your primary agenda, however, is to dissipate the energy generated by the novelty of leaving the classroom during class time. Focus on preparing your transition to the outdoors, moving out in a purposeful and orderly fashion and arriving at your predetermined "home base," a destination that you will always go to initially when you leave the classroom. Form a sharing circle, a process you will use time and again. Tour the schoolyard, proceeding as a whole group, then ask students to walk as individuals for a few minutes, then with a partner for a minute, and finally with a group of four. This gives students a brief experience with each of the four ways they will be organized for various outdoor activities. End with another sharing circle and an orderly return to the classroom.

**Challenging students.** Sometimes the class will be inattentive and unresponsive. At such times, it is appropriate to direct students back to the classroom. Breaking the contract has consequences, and students need to understand that the opportunity to learn in nature is a privilege. They will remember that day.

In rare instances, you may have an individual student who is regularly not able to comply. Interestingly, this is probably not the student who you anticipated would have difficulty. Often, students who have difficulty with attention and performance in traditional classroom seat work shine and take leadership outdoors. In the case of the noncompliant student, it may be necessary to ask him or her to take a time-out if he or she is able to sit without disrupting others' experiences. If the bad behavior persists, you may be obliged to return to the classroom early. Even so, it is important to give the student a chance to redeem himself or herself the next time you go outdoors.

## Routines

Routines are good for management. They impose a measure of self-monitoring because they represent behaviors that are already known and have been practiced. If one person transgresses during a routine, other students are able to intervene to help you with student management. Here are a few routines that may work for you.

**Science door.** Have you ever watched a group of students pass through the exterior door on their way to recess? As soon as one foot hits the asphalt, they start running and cheering. It is a beautiful sight. Clearly, this is not how you want students to leave the building as you head out for science. One subtle but effective way to distinguish science from recess is to use a different door for science than for recess. Refer to this exit as the "learning door."

**Transition behavior.** Be explicit about how you want students to walk through the hallways and into the schoolyard. If students exit wildly, simply ring the bell, have them line up inside, and try it again. If this continues to be a problem, return to the classroom and try another day. By consistently showing students that this behavior limits their time outdoors, they will follow your directions.

**Home base.** Establish a destination in the schoolyard where every outdoor activity will begin. Students should walk directly there after leaving the school building. Choose a place that is level and, if possible, away from classroom windows and popular recess areas.

**Sharing circle.** When students arrive at home base, they should form a large sharing circle—everyone in a single ring with no students hanging back. This is an effective way to maintain eye contact with all students while you give instructions or share findings. Take a position in the circle where you are facing the Sun. This way, you will know that students won't be distracted by having the Sun's glare in their eyes. A sharing circle is also used to transition from one task to another, to summarize an activity, or anytime you need to regroup.

Techniques for forming a circle vary. One method is "magnetic feet." Students spread their legs to meet their neighbors' feet. Magically, these magnets turn off when you direct them to do so. Students may also stand with hands on hips, elbows touching with neighbors'. Pick or create a method that works with your students.

To speed up the formation of a sharing circle, try the tried-and-true countdown from five, with the objective that everyone is in a proper circle by zero.

**Attention signal.** Adopt a uniform signal for attention. It is essential that students respond to the attention signal immediately. You may choose the same method you use in the classroom or, if this is not appropriate for the outdoors, try one of these.

- A chime, bell, whistle, or other singular and loud sound. These are appropriate outdoors. When students hear it, they stop, look, and listen.

- Count down from five and when you get to one, students are silent with their hands up. This might not be appropriate outdoors. A countdown from ten can be used to call students back to a sharing circle.

- Clap call and response. You clap a pattern, and students return it by repeating the clapping pattern. This works if students are all nearby.

**Focus question.** Inquiry-based activities are guided by a question. This pedagogical routine should extend into the schoolyard, too. *Students need to know why they are engaged in the outdoor investigation.* They should expect to write the focus question in their notebooks at the outset of the investigation and produce an answer at the end of the investigation.

**Boundaries.** Setting boundaries allows students freedom within a defined space. Because different activities may require different locations, it is always important to be explicit about where students are allowed to travel during the outdoor activity.

**Buddy system.** You may want to institute a buddy system, particularly if you leave the schoolyard. When participants are paired off, tell them that each individual is responsible at all times for the whereabouts and safety of his or her buddy. It is helpful and fun to number the pairs in order to count off quickly and account for everyone.

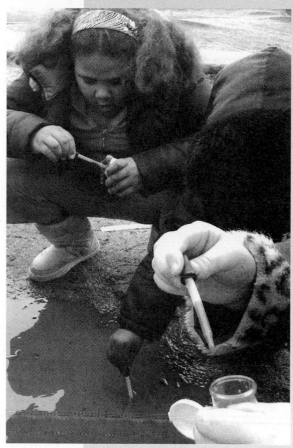

## Considerations for Students with Disabilities

FOSS evolved from pioneering work done in the 1970s with students with physical disabilities. The legacy of that work is that FOSS investigations incorporate multisensory methods, not only to accommodate students with physical and learning disabilities, but also to maximize information gathering for all students. Strategies that provide opportunities to learn for students with disabilities turn out to be good strategies for all students.

All students benefit from opportunities to experience the natural world outdoors. For students with disabilities, consider how to make the schoolyard accessible and safe so that they can work with a degree of independence. This requires advance planning to make sure that the student, his or her family, the special education teacher, and others involved in the child's school experience are informed and have input into the process.

Whenever a student with a disability is successful in a full-inclusion classroom, there is a behind-the-scenes collaborative effort of caring educators who work together to support the student with just the right amount of scaffolding. In advance of teaching the first outdoor activity, contact the special education teachers in charge of each student's Individualized Education Plan (IEP), and have them review the planned outdoor experiences. Ask the teachers to recommend modifications that will better accommodate and support each student. Invite the special educators to join the class for the outdoor activity.

# Taking FOSS Outdoors

***Attention and language-based disabilities.*** Students with attention and behavioral issues often thrive when they are engaged in science outdoors. A fenced area will help you to both keep track of students and provide a sense of safety. Having students work with partners (buddies) allows students to look after each other. Provide short, structured opportunities for students to participate in outdoor activities in a clearly defined space, and expand the boundaries and time expectations as students earn your trust and confidence. Many educators have found great success with treating the outdoor activity as a reward for excellent behavior indoors.

Consider students' communication requirements, and plan to bring specialty devices outdoors with you. This might be as simple as some picture cue cards to help enhance your message or an electronic communication device such as a computer.

***Physical disabilities and visual impairments.*** In the Getting Ready section of each outdoor activity, we ask that the teacher decide where the outdoor activity should be taught. You may find that certain locations are better than others for the purposes of developing science concepts and meeting the physical needs of students with disabilities. Get to know your schoolyard really well, and try to experience it as your students do.

One side of most schoolyards is typically a parking lot, and the other three sides have spaces accessible to students for work and play. If you have a student with a physical disability, you need to consider if the terrain provides for good mobility for the student. Often, schoolyards are handicapped accessible because they are covered in asphalt. For many of the FOSS outdoor activities, an asphalt area is appropriate to use. When you want to use a greener location, make sure wheelchairs or crutches will work on these new surfaces. If the surface will be a mobility challenge, see if a paraprofessional or an educational assistant is able to help the student. If someone is not able to join you, consider if a classmate can help. If this is not an option, then consider working at a transition zone where the grass meets the asphalt.

A student with a visual impairment should make a scouting trip to the outdoor site with a mobility instructor to get the lay of the land and to learn where things are located. If the student becomes familiar with and knows how to navigate his or her outdoor surroundings, it will allow for more independence. Even so, during the actual outdoor activity, the student may need someone to quietly describe the terrain ahead and may need a fellow student's arm for balance and security.

If a student struggles with gross motor coordination, uneven ground may present a challenge. Just as you would in the classroom, begin by offering more support, and slowly pull back on this assistance as students become more comfortable with their stamina, security, and endurance with regular outdoor activities.

***Sensory sensitivity.*** For a student with tactile-sensitivity issues, make it clear that he or she may, for example, observe as a classmate digs in the soil to collect a sample. Over time, this student may feel better able to participate by using gloves or by washing his or her hands as soon as the digging is complete. Knowing where each student falls on the continuum of a disability will help you decide when to hold back asking a student to fully participate, when to allow him or her to just observe, and when to give a gentle nudge and expect more active participation.

For students with sensory disorders, the outdoors is often a calming space. Consider where the quietest place in the schoolyard is, and use this more often if you have students with sensitivity to noise.

No matter what the disability, educators have found success taking students outdoors. With advance planning, communication with the student and the special education team, and a little extra effort, you, too, can provide a rich, safe outdoor learning experience for all your students.

## TEACHING STRATEGIES

In the beginning, you may find that students regularly use descriptive terms such as "icky," "yucky," and "gross." You may have students who say things such as "I cannot get my clothes dirty. My mom will be mad." Many students are fearful of bugs, wooded areas, and even just sitting on the grass. Often, after a few outdoor activities, these fears and excuses fade away. With patience, persistence, and support, students' resistance may be overcome entirely. If you suspect that your students may be reluctant to work outdoors, structure your first few activities to be low-stress activities. The first few times outdoors can be fairly benign activities with students choosing a comfortable place to just sit

(or stand), practicing writing outdoors, and doing simple collecting or counting tasks.

**Set the tone.** Many teaching strategies that are effective in the classroom work outdoors, too. For example, at the sharing circle, instead of instinctively talking louder (because it is noisy outdoors), kneel down and speak in a loud whisper so that students need to focus to hear you. If students are speaking, put up your silent signal, and wait for silence. The educator's voice sets the tone for the activity.

**Take a position.** In the sharing circle, position yourself where you have the Sun in your face so that students don't need to squint. If possible, place yourself next to those students who might benefit from a silent look or hand on the shoulder to remind them to be silent.

**Meet the challenge.** Students who struggle with behavior problems often respond well outdoors when given responsibility. Let the active student carry the heavy jug of water or take the position at the front of the line to lead the class outdoors. For many students, this is all it takes to get them off on the right foot for the outdoor activity.

Students who have the greatest difficulty controlling their behavior indoors are often the leaders when it comes to working in an outdoor space. You may find that students who are not as attentive or cannot sit still inside are the most insistent about quieting down so that the class can get outdoors for science.

**Get them writing.** Primary students (grades K–2) can fill out a chart on a clipboard outdoors. They are also capable of recording observations outdoors in their notebooks if observation is their only task. Most primary students will need to sit down with their clipboards on their laps or on the ground to do this successfully. In the early years, most writing follows an outdoor activity and is done inside on desks and with the classroom's word wall.

Upper-elementary students (grades 3–5) are capable of writing outdoors. Students will benefit from a quick activity about how to place the notebook or clipboard in the crook of their nonwriting arm for support.

Depending on the activity, you might decide to have students attach their notebooks to a clipboard and place the clipboards in a crate for easy transport and storage. This technique is useful when the ground is moist, when the activity is messy, or when students need to use their hands to complete the activity. The recording will happen immediately after the hands-on activity. Be open to the surprise of how much your students are capable of noticing and recording during and after an outdoor activity.

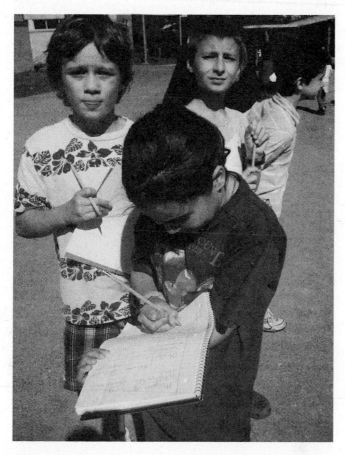

# FLOW OF OUTDOOR ACTIVITIES

The natural flow of a FOSS outdoor activity is slightly different from that of a standard FOSS indoor activity. The steps of a typical outdoor activity are listed below. This list may be helpful if you want to teach more than the handful of outdoor activities in the *Investigations Guide*, or if you want to adapt an indoor activity for schoolyard use.

1. **Prepare for the outdoor activity.**

   • Determine the best location to teach the activity.

   • Check the weather forecast.

   • Make sure students will be dressed appropriately.

   • Prepare materials for distribution.

   • Check the site the morning of the activity.

2. **Set the learning objective.**

   • Present the focus question.

   • Discuss procedures.

3. **Go outdoors.**

   • Gather at the predetermined location.

4. **Model or describe the activity.**

   • Organize students.

   • Define boundaries.

   • Introduce/distribute materials.

5. **Monitor the activity.**

   • Check student engagement.

   • Check student recording.

   • Ask questions.

6. **Share the experience.**

   • Form a sharing circle to discuss experiences.

   • Share thinking.

   • Share answers to the focus question.

7. **Return to class**

   • Make connections to the related indoor activity.

   • Display student work and collections.

# EXTENDING BEYOND FOSS OUTDOOR ACTIVITIES

Occasionally, you may stumble upon a serendipitous opportunity. A breeze may launch thousands of twirling seeds from a maple tree, a woodpecker may alight on a tree so close that students can observe it drumming for insects, student-made parachutes may be carried by an updraft high into the sky and out of sight. To your delight, you may spy something you have never seen before. It can happen at any time when you are outdoors!

At special moments like these, our job as educators is to signal students to stop and quietly appreciate the suspension of time. Sometimes, words break the wonder. Trust your instincts at magical moments like these. The answers to questions will come eventually. It is not essential to label the event or even understand it. By inviting students to be alive with their feelings in the moment, you give them a gift for a lifetime.

It is not uncommon for educators to experience the powerful effect of the outdoors on student learning. If you find yourself searching for other outdoor learning opportunities, consider the ones below.

***Move activities outdoors.*** Whatever the subject, students will have more room outdoors to be creative with some activities, and you can worry less about water, sand, and gravel spills. You must still consider how to transport materials, where students will sit, how they will return their project to the classroom, and how to clean up the outdoor space and students' hands before returning to the building.

***Use the outdoors for extensions.*** Extending an inside concept to the outdoors is an excellent way to apply new knowledge. For example, in the **Structures of Life Module**, students grow bush beans hydroponically. If the large leaves fascinate students, go outdoors and see how many kinds of leaves you can find in the schoolyard. Do they all have smooth edges and come to a point at their tips? Go on a leaf hunt, group the leaves by their characteristics, and, eventually, have students tape them into their science notebooks.

There is great value in repeating an indoor activity outdoors. If your students are sanding wood samples inside, follow this up with a trip outdoors to find a stick and sand it. Have you been studying sow bugs? Ask students if they think they know where in the schoolyard they might find these bugs. Applying what students have learned in the classroom and putting that knowledge to work outdoors is an effective way to solidify their understanding. It's also an effective way to informally assess whether students understand the concepts, as well as a method for reinforcing the learning.

**Find solitude.** Use your outdoor space for silent independent work time. Just as in the classroom, the outdoor space can be a workspace with activities going on. At times, the outdoor space is more of a sanctuary for independent observation and notebook writing. It can be a place for special classroom rituals, awakening awareness of the beauty of nature. Sometimes, it can just be a place to be silent for a minute to awaken the senses and refocus students' energy. Some teachers increase this silent minute to 2, 3, or even 5 minutes. Silence is something to be practiced, and for many students and teachers, this can be challenging. This is a special way to end an outdoor experience and will help students transition into the classroom.

**Enhance biodiversity.** Modify your schoolyard by adding natural materials, such as logs, rocks, or paving stones. These structures can provide safe havens that may attract more living things. These types of shelters can be particularly helpful if you have an environment without natural shelter from the Sun, such as trees and shrubs. Students can also be involved in the design and implementation of these projects.

Schoolyard modification of this kind requires administrative participation and the support of the school custodian. Marking the area with educational signage can further benefit the enhanced site. If your schoolyard habitat needs your intervention to cultivate biodiversity, understand that it can take a couple of years to get established. Areas completely surrounded by blacktop or concrete can become filled with living things if provided with food, shelter, and water.

**Attract wildlife.** There are many responsible ways to attract wildlife to your class windows with feeders for birds, squirrels, hummingbirds, or butterflies, as well as many great programs for monitoring these animals. See FOSSweb for ideas for additional wildlife observation projects.

**Establish long-term studies.** The possibilities for long-term studies are endless, ranging from weather monitoring to seasonal population variation. It can be as simple as adopting an observation location and visiting it monthly to monitor various aspects of change over time. See FOSSweb for ideas for long-term projects.

**Create gardens.** Planting a garden in raised beds or improved soil is an ambitious option for increasing the biodiversity of your schoolyard. Consider carefully, especially with a vegetable garden, the timing of the school year. In most parts of the country, the time when plants require the most support is during summer vacation. Even if you can get a summer program involved, we suggest starting with indigenous plants that bloom or mature in spring and fall and require little maintenance.

▶ **NOTE**

Remember not to release any classroom animals into schoolyard environments.

# ELEMENTARY-LEVEL ENVIRONMENTAL EDUCATION

In the early 1990s, David Sobel noticed something poignant about children's perceptions of the environment. If a child had been introduced to environmental issues at school that were presented in the context of doom-and-gloom scenarios, the child expressed a heightened sense of anxiety and hopelessness, which Sobel calls ecophobia (Sobel 1996). The implications of his finding should raise a cautionary flag. Sobel is not suggesting that we abandon teaching about the environment in our elementary schools. He is proposing a different approach to environmental education that will bring our children into natural, healthy relationships with environmental issues.

Effective early environmental education should focus on local and ultralocal issues. What is happening in our schoolyards? What factors influence the communities of plants and animals in our neighborhoods? How do changing weather conditions affect the populations around our schools? How are our actions affecting the habitats in our schoolyards? What can we do to enhance natural systems at our schools? Elaborate rain forest projects provide little understanding and have negligible impact on students' connections to nature; researching and installing a butterfly garden or keeping an inventory of the birds in the schoolyard can be transformative. The children from Sobel's 1996 study could tell you how many species in the Amazon were going extinct each minute, but were unfamiliar with the most common plants in their schoolyard.

Time outdoors during the school day is beneficial for student learning. Students who are exposed to hands-on experiences in their local environment often become enthusiastic, self-motivated learners and, typically, academically outperform their peers who do not have these learning opportunities (Liebermann and Hoody 1998). Children are able to pay attention for longer periods of time outdoors on the same assignment and are more focused when they return to their indoor class work (Louv 2008).

Research has produced evidence that using the schoolyard is an effective way to enhance student learning. Texas A&M University, in conjunction with the Texas Education Agency, conducted a meta-analysis of the research in order to identify and rank effective instructional methods for science education and to define how best to improve student achievement. The highest-ranked teaching strategy was Enhanced Context Strategies, which included taking meaningful field trips and using the schoolyard for activities (Scott et al. 2005).

Students' attitudes toward learning are influenced by simple outdoor experiences. In one study (Shaw and Terrance 1981), students who experienced outdoor instruction reported that, in general, they enjoyed school more and felt more supported and trusted by their teacher than they had prior to the outdoor experiences. These pretest/posttest differences were more pronounced for students who had been identified as being "uninvolved" in the classroom activities. Also, this student perception was a lasting effect that carried over to the regular classroom activities weeks later.

Perhaps the most important benefit of incorporating the outdoors into the traditional school learning environment is that it offers opportunities for students to synthesize concepts and personal experience by applying what they have learned to a new environment.

FOSS outdoor activities will help you focus on age-appropriate environmental topics and enable you to create meaningful and personal connections between your students and their local environment. When students can openly explore the environment, they can create meaningful connections to their learning and establish positive relationships with nature. You'll be amazed by what students notice.

Here's the good news. If you focus on inquiry and direct experience instead of problems, it takes remarkably little guidance for students to make positive, empowering, lifelong connections to nature. One insightful young man explained, "My video games have a pattern that is always the same, but nature is like a game that is different every time you play." As an educator, you can draw out that sense of wonder and curiosity for students while simultaneously helping them build a solid science foundation.

# REFERENCES

Liebermann, G., and L. Hoody. 1998. *Closing the Achievement Gap: Using the Environment as an Integrated Context for Learning; Results of a National Study.* San Diego: State Education and Environment Roundtable.

Louv, R. 2008. *Last Child in the Woods: Saving Our Children from Nature-Deficit Disorder.* New York: Workman Publishing.

Scott, T. et al. 2005. *Texas Science Initiative Meta-Analysis of National Research Regarding Science Teaching.* Texas Education Agency.

Shaw, T. J., and J. M. Terrance. 1981. "Involved and Uninvolved Student Perceptions in Indoor and Outdoor School Settings." *Journal of Early Adolescence,* 1:135–146.

Sobel, D. 1996. *Beyond Ecophobia: Reclaiming the Heart in Nature Education.* Great Barrington, MA: Orion Society.

## ACKNOWLEDGMENTS

The Taking FOSS Outdoors initiative got its start through a collaboration with the Boston Schoolyard Initiative (BSI). In 2004, BSI began developing an approach to teaching science that routinely takes students into the schoolyard to test, apply, and explore core science concepts and skills. As part of this project, BSI developed *Science in the Schoolyard Guides*™ for 12 FOSS modules and a companion *Science in the Schoolyard*™ DVD. In partnership with the City of Boston, BSI designs and builds schoolyards that provide a rich environment for teaching, learning, and play. For more information on BSI, *Science in the Schoolyard*, or BSI's *Outdoor Writer's Workshop*™ professional development program and materials, see www.schoolyards.org.

# Science Notebook Masters

# Mealworm Observations

1. Observe the structures of a mealworm.

2. Draw a mealworm and label its structures (parts of body).

3. List three mealworm structures. Describe each one's function (what the structure helps the mealworm do).

4. Observe and describe the mealworm's behavior (what it does).

5. What questions do you have about mealworms?

6. What do you need to know about the mealworm to keep it alive and healthy in the classroom?

FOSS Next Generation
© The Regents of the University of California
Can be duplicated for classroom or workshop use.

Environments Module
Investigation 1: Environmental Factors
No. 1—Notebook Master

# Mealworm Observations

1. Observe the structures of a mealworm.

2. Draw a mealworm and label its structures (parts of body).

3. List three mealworm structures. Describe each one's function (what the structure helps the mealworm do).

4. Observe and describe the mealworm's behavior (what it does).

5. What questions do you have about mealworms?

6. What do you need to know about the mealworm to keep it alive and healthy in the classroom?

# Mealworm Observations

1. Observe the structures of a mealworm.

2. Draw a mealworm and label its structures (parts of body).

3. List three mealworm structures. Describe each one's function (what the structure helps the mealworm do).

4. Observe and describe the mealworm's behavior (what it does).

5. What questions do you have about mealworms?

6. What do you need to know about the mealworm to keep it alive and healthy in the classroom?

FOSS Next Generation
© The Regents of the University of California
Can be duplicated for classroom or workshop use.

Environments Module
Investigation 1: Environmental Factors
No. 1—Notebook Master

# Mealworm Observations

1. Observe the structures of a mealworm.

2. Draw a mealworm and label its structures (parts of body).

3. List three mealworm structures. Describe each one's function (what the structure helps the mealworm do).

4. Observe and describe the mealworm's behavior (what it does).

5. What questions do you have about mealworms?

6. What do you need to know about the mealworm to keep it alive and healthy in the classroom?

# Life of a Mealworm

| | Observation 1 | Observation 2 | Comments |
|---|---|---|---|
| Week 1 | | | |
| Week 2 | | | |
| Week 3 | | | |
| Week 4 | | | |

Environments Module
Investigation 1: Environmental Factors
No. 2—Notebook Master

FOSS Next Generation
© The Regents of the University of California
Can be duplicated for classroom or workshop use.

# Life of a Mealworm

| | Observation 1 | Observation 2 | Comments |
|---|---|---|---|
| Week 5 | | | |
| Week 6 | | | |
| Week 7 | | | |
| Week 8 | | | |

Environments Module
Investigation 1: Environmental Factors
No. 2—Notebook Master

FOSS Next Generation
© The Regents of the University of California
Can be duplicated for classroom or workshop use.

# Isopod Investigation

**Part 1. Setup**

Number of pill bugs _____    Number of sow bugs _____

Environmental factor tested _____

Describe how you set up the container. _____

**Part 2. Observations**

Record where each animal was and what it was doing (on surface, buried, moving).

This is where the animals were after _____ minutes.

**Short run**

This is where the animals were after _____ hours.

**Long run**

**Part 3. Interpretation**

What can you say about the animals' environmental preferences?

FOSS Next Generation
© The Regents of the University of California
Can be duplicated for classroom or workshop use.

Environments Module
Investigation 1: Environmental Factors
No. 3—Notebook Master

---

# Isopod Investigation

**Part 1. Setup**

Number of pill bugs _____    Number of sow bugs _____

Environmental factor tested _____

Describe how you set up the container. _____

**Part 2. Observations**

Record where each animal was and what it was doing (on surface, buried, moving).

This is where the animals were after _____ minutes.

**Short run**

This is where the animals were after _____ hours.

**Long run**

**Part 3. Interpretation**

What can you say about the animals' environmental preferences?

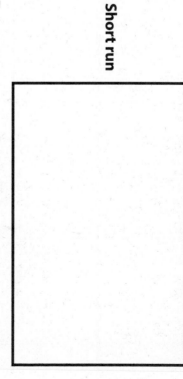

FOSS Next Generation
© The Regents of the University of California
Can be duplicated for classroom or workshop use.

Environments Module
Investigation 1: Environmental Factors
No. 3—Notebook Master

# Isopod Environment Map

Key

1. How much water did you put into your terrarium?

2. Where did you put it?

3. List the environmental factors.

FOSS Next Generation
© The Regents of the University of California
Can be duplicated for classroom or workshop use.

Environments Module
Investigation 1: Environmental Factors
No. 4—Notebook Master

# Isopod Environment Map

Key

1. How much water did you put into your terrarium?

2. Where did you put it?

3. List the environmental factors.

FOSS Next Generation
© The Regents of the University of California
Can be duplicated for classroom or workshop use.

Environments Module
Investigation 1: Environmental Factors
No. 4—Notebook Master

# Response Sheet—Investigation 1

A student wanted to find out isopod preferences for temperature. Below is his notebook entry.

We put dry, moist, and wet soil in the basins, just like we did before in class. Next, we put a heating pad under part of the basin to make three areas, each with a different temperature. That way we could find out if isopods like it cold, warm, or hot.

| Cold area | Warm area | Hot area |
|---|---|---|
| Dry soil | Moist soil | Wet soil |

Results

After 10 minutes, we found some isopods in the warm and hot areas. After an hour, we found all of the isopods in the warm area. Inference: Isopods like it warm.

1. Do you agree that the student's investigation supports his inference? Why or why not?

2. What would you do differently to improve his investigation?

FOSS Next Generation
© The Regents of the University of California
Can be duplicated for classroom or workshop use.

Environments Module
Investigation 1: Environmental Factors
No. 5—Notebook Master

# Response Sheet—Investigation 1

A student wanted to find out isopod preferences for temperature. Below is his notebook entry.

We put dry, moist, and wet soil in the basins, just like we did before in class. Next, we put a heating pad under part of the basin to make three areas, each with a different temperature. That way we could find out if isopods like it cold, warm, or hot.

| Cold area | Warm area | Hot area |
|---|---|---|
| Dry soil | Moist soil | Wet soil |

Results

After 10 minutes, we found some isopods in the warm and hot areas. After an hour, we found all of the isopods in the warm area. Inference: Isopods like it warm.

1. Do you agree that the student's investigation supports his inference? Why or why not?

2. What would you do differently to improve his investigation?

FOSS Next Generation
© The Regents of the University of California
Can be duplicated for classroom or workshop use.

Environments Module
Investigation 1: Environmental Factors
No. 5—Notebook Master

## Critter Record

Your name: _____

Date: _____

Use the Critter Replicator to help you draw your selected organism.

[ ]

Head    Thorax    Abdomen

Size:    Draw a line as long as the organism.

Where did you find your organism?

_____
_____
_____

How did your organism behave?

_____
_____
_____

FOSS Next Generation
© The Regents of the University of California
Can be duplicated for classroom or workshop use.

Environments Module
Investigation 1: Environmental Factors
No. 6—Notebook Master

## Critter Record

Your name: _____

Date: _____

Use the Critter Replicator to help you draw your selected organism.

[ ]

Head    Thorax    Abdomen

Size:    Draw a line as long as the organism.

Where did you find your organism?

_____
_____
_____

How did your organism behave?

_____
_____
_____

FOSS Next Generation
© The Regents of the University of California
Can be duplicated for classroom or workshop use.

Environments Module
Investigation 1: Environmental Factors
No. 6—Notebook Master

## Critter Body Parts

**HEAD**

**THORAX**

**ABDOMEN**

**WINGS**

---

## Critter Body Parts

**HEAD**

**THORAX**

**ABDOMEN**

**WINGS**

# Living and Nonliving Factors

| | |
|---|---|
| salamanders | L |
| cricket | L |
| rocks | |
| pan of water | |
| broad-leaved plants | L |
| light | L |
| grassy plants | L |
| glass terrarium | |
| thin-leaved plants | L |
| soil | L |
| flowering plants | L |
| temperature | |

A student used the picture above to make a list of all the environmental factors she saw in this terrestrial environment. She put an *L* next to each factor she identified as a living factor. The list the student made appears above.

1. Do you agree that all the factors she marked with an *L* are living? If not, tell which ones you don't agree with and explain why you don't agree.

2. Describe how two of the nonliving factors might influence the living factors in this terrarium.

FOSS Next Generation
© The Regents of the University of California
Can be duplicated for classroom or workshop use.

Environments Module
Investigation 2: Ecosystems
No. 8—Notebook Master

# Living and Nonliving Factors

| | |
|---|---|
| salamanders | L |
| cricket | L |
| rocks | |
| pan of water | |
| broad-leaved plants | L |
| light | L |
| grassy plants | L |
| glass terrarium | |
| thin-leaved plants | L |
| soil | L |
| flowering plants | L |
| temperature | |

A student used the picture above to make a list of all the environmental factors she saw in this terrestrial environment. She put an *L* next to each factor she identified as a living factor. The list the student made appears above.

1. Do you agree that all the factors she marked with an *L* are living? If not, tell which ones you don't agree with and explain why you don't agree.

2. Describe how two of the nonliving factors might influence the living factors in this terrarium.

FOSS Next Generation
© The Regents of the University of California
Can be duplicated for classroom or workshop use.

Environments Module
Investigation 2: Ecosystems
No. 8—Notebook Master

| Week | Date | Time | Water temp. | Observations |
|---|---|---|---|---|
| 1 | | | | |
| 2 | | | | |
| 3 | | | | |
| 4 | | | | |
| 5 | | | | |
| 6 | | | | |
| 7 | | | | |

| Week | Date | Time | Water temp. | Observations |
|---|---|---|---|---|
| 1 | | | | |
| 2 | | | | |
| 3 | | | | |
| 4 | | | | |
| 5 | | | | |
| 6 | | | | |
| 7 | | | | |

## Practice with Food Chains

Make several food chains of at least three organisms. Use arrows to show how food moves from organism to organism.

- American robin
- Aquatic snail
- Bacteria
- Black bear
- Brook trout
- Chipmunk
- Coyote
- Dead plants and animals
- Earthworm
- Grama grass
- Great blue heron
- Green algae
- Grouse
- Hare
- Mayfly
- Pine trees
- Red-tailed hawk
- Scuds
- *Tubifex* worm
- Wild blueberry

FOSS Next Generation
© The Regents of the University of California
Can be duplicated for classroom or workshop use.

Environments Module
Investigation 2: Ecosystems
No. 10—Notebook Master

## Practice with Food Chains

Make several food chains of at least three organisms. Use arrows to show how food moves from organism to organism.

- American robin
- Aquatic snail
- Bacteria
- Black bear
- Brook trout
- Chipmunk
- Coyote
- Dead plants and animals
- Earthworm
- Grama grass
- Great blue heron
- Green algae
- Grouse
- Hare
- Mayfly
- Pine trees
- Red-tailed hawk
- Scuds
- *Tubifex* worm
- Wild blueberry

FOSS Next Generation
© The Regents of the University of California
Can be duplicated for classroom or workshop use.

Environments Module
Investigation 2: Ecosystems
No. 10—Notebook Master

# Response Sheet—Investigation 2

Here is a description of the organisms that live in a pond.

Aquatic snails and mayflies eat algae.

Frogs and trout eat mayflies.

Herons eat frogs, trout, and snails.

Bacteria decompose the dead organisms.

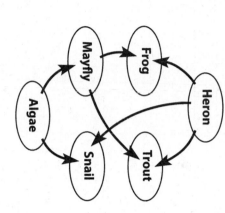

A student read the information about organisms living in a pond. Then she drew a food web.

Is this food web accurate? You can draw a new food web if needed.

Write your comments and draw on the facing blank notebook page.

FOSS Next Generation
© The Regents of the University of California
Can be duplicated for classroom or workshop use.

Environments Module
Investigation 2: Ecosystems
No. 11—Notebook Master

---

# Response Sheet—Investigation 2

Here is a description of the organisms that live in a pond.

Aquatic snails and mayflies eat algae.

Frogs and trout eat mayflies.

Herons eat frogs, trout, and snails.

Bacteria decompose the dead organisms.

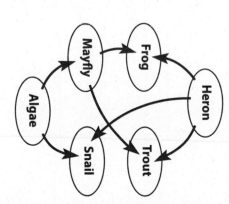

A student read the information about organisms living in a pond. Then she drew a food web.

Is this food web accurate? You can draw a new food web if needed.

Write your comments and draw on the facing blank notebook page.

FOSS Next Generation
© The Regents of the University of California
Can be duplicated for classroom or workshop use.

Environments Module
Investigation 2: Ecosystems
No. 11—Notebook Master

# Population Simulation Results

| Year | # of deer at start | # of deer at end | Survival predictions |
|---|---|---|---|
| 1 | | | |
| 2 | | | |
| 3 | | | |
| 4 | | | |
| 5 | | | |

In year 2, we started with _____ deer and ended with _____ .

The number of deer _____ because _____

_____

_____

In year 4, we started with _____ deer and ended with _____ .

The number of deer _____ because _____

_____

_____

In year 5, we started with _____ deer and ended with _____ .

The number of deer _____ because _____

_____

_____

FOSS Next Generation
© The Regents of the University of California
Can be duplicated for classroom or workshop use.

Environments Module
Investigation 2: Ecosystems
No. 12—Notebook Master

# Population Simulation Results

| Year | # of deer at start | # of deer at end | Survival predictions |
|---|---|---|---|
| 1 | | | |
| 2 | | | |
| 3 | | | |
| 4 | | | |
| 5 | | | |

In year 2, we started with _____ deer and ended with _____ .

The number of deer _____ because _____

_____

_____

In year 4, we started with _____ deer and ended with _____ .

The number of deer _____ because _____

_____

_____

In year 5, we started with _____ deer and ended with _____ .

The number of deer _____ because _____

_____

_____

FOSS Next Generation
© The Regents of the University of California
Can be duplicated for classroom or workshop use.

Environments Module
Investigation 2: Ecosystems
No. 12—Notebook Master

# RESPONSE SHEET—INVESTIGATION 2

A boy told his friend he heard frogs in a pond the previous night. He said, "I think the frogs were croaking because it was a nice warm night and they were happy."

The friend said, "I don't think they were croaking because they were happy. I think they were communicating with one another."

What do you think?

FOSS Next Generation
© The Regents of the University of California
Can be duplicated for classroom or workshop use.

Environments Module
Investigation 2: Ecosystems
No. 13—Notebook Master

# RESPONSE SHEET—INVESTIGATION 2

A boy told his friend he heard frogs in a pond the previous night. He said, "I think the frogs were croaking because it was a nice warm night and they were happy."

The friend said, "I don't think they were croaking because they were happy. I think they were communicating with one another."

What do you think?

FOSS Next Generation
© The Regents of the University of California
Can be duplicated for classroom or workshop use.

Environments Module
Investigation 2: Ecosystems
No. 13—Notebook Master

# RESPONSE SHEET—INVESTIGATION 2

A boy told his friend he heard frogs in a pond the previous night. He said, "I think the frogs were croaking because it was a nice warm night and they were happy."

The friend said, "I don't think they were croaking because they were happy. I think they were communicating with one another."

What do you think?

# RESPONSE SHEET—INVESTIGATION 2

A boy told his friend he heard frogs in a pond the previous night. He said, "I think the frogs were croaking because it was a nice warm night and they were happy."

The friend said, "I don't think they were croaking because they were happy. I think they were communicating with one another."

What do you think?

# Brine Shrimp–Hatching Experiment

## Question

Write the question you will answer with your investigation.

## Materials

| | |
|---|---|
| 4 | Plastic cups with lids |
| 4 | Labels |
| 1 | Cup of salt |
| 1 | Spoon, 5 mL |
| 1 | Beaker, 100 mL |
| • | Tape |
| 1 | Vial of brine shrimp eggs |
| 1 | Minispoon |
| 1 | Craft stick |
| 1 | FOSS tray |
| 1 | Container of water |

## Procedure

1. Label four cups—0 spoons, 2 spoons, 4 spoons, and 6 spoons salt. Write your group name on each label.

2. Measure 150 mL of water into each cup. Use the 100 mL beaker.

3. Measure salt into the labeled cups, using the 5 mL spoon. Use a craft stick to level the measure. Put lids on the cups. Gently swirl the cups to dissolve the salt.

4. Put 1 level minispoon of brine shrimp eggs in each cup. Put the lids on. Gently swirl the water to wet the eggs.

5. Place all four cups on the FOSS tray.

FOSS Next Generation
© The Regents of the University of California
Can be duplicated for classroom or workshop use.

Environments Module
Investigation 3: Brine Shrimp Hatching
No. 14—Notebook Master

**WARNING** — This set contains chemicals that may be harmful if misused. Read cautions on individual containers carefully. Not to be used by children except under adult supervision.

# Brine Shrimp–Hatching Experiment

## Question

Write the question you will answer with your investigation.

## Materials

| | |
|---|---|
| 4 | Plastic cups with lids |
| 4 | Labels |
| 1 | Cup of salt |
| 1 | Spoon, 5 mL |
| 1 | Beaker, 100 mL |
| • | Tape |
| 1 | Vial of brine shrimp eggs |
| 1 | Minispoon |
| 1 | Craft stick |
| 1 | FOSS tray |
| 1 | Container of water |

## Procedure

1. Label four cups—0 spoons, 2 spoons, 4 spoons, and 6 spoons salt. Write your group name on each label.

2. Measure 150 mL of water into each cup. Use the 100 mL beaker.

3. Measure salt into the labeled cups, using the 5 mL spoon. Use a craft stick to level the measure. Put lids on the cups. Gently swirl the cups to dissolve the salt.

4. Put 1 level minispoon of brine shrimp eggs in each cup. Put the lids on. Gently swirl the water to wet the eggs.

5. Place all four cups on the FOSS tray.

FOSS Next Generation
© The Regents of the University of California
Can be duplicated for classroom or workshop use.

Environments Module
Investigation 3: Brine Shrimp Hatching
No. 14—Notebook Master

## Brine Shrimp–Hatching Observations

### Part 1

What day of your experiment did you first notice hatching? (Count setup day as day 1.)

☐ Day 1  ☐ Day 2  ☐ Day 3  ☐ Day 4

In which container did you first observe hatching?

☐ 0 spoons  ☐ 4 spoons
☐ 2 spoons  ☐ 6 spoons

What additional observations did you make? Write your observations on the next blank page in your notebook.

### Part 2

Record the number of eggs that hatched in each salt condition.

Make an X in one box for each salt condition.

Day ——— (Count setup day as day 1.)

**Salt Conditions**

|  | 0 spoons | 2 spoons | 4 spoons | 6 spoons |
|---|---|---|---|---|
| Most |  |  |  |  |
| Some |  |  |  |  |
| None |  |  |  |  |

FOSS Next Generation
© The Regents of the University of California
Can be duplicated for classroom or workshop use.

Environments Module
Investigation 3: Brine Shrimp Hatching
No. 15—Notebook Master

**WARNING —** This set contains chemicals that may be harmful if misused. Read cautions on individual containers carefully. Not to be used by children except under adult supervision.

## Brine Shrimp–Hatching Observations

### Part 1

What day of your experiment did you first notice hatching? (Count setup day as day 1.)

☐ Day 1  ☐ Day 2  ☐ Day 3  ☐ Day 4

In which container did you first observe hatching?

☐ 0 spoons  ☐ 4 spoons
☐ 2 spoons  ☐ 6 spoons

What additional observations did you make? Write your observations on the next blank page in your notebook.

### Part 2

Record the number of eggs that hatched in each salt condition.

Make an X in one box for each salt condition.

Day ——— (Count setup day as day 1.)

**Salt Conditions**

|  | 0 spoons | 2 spoons | 4 spoons | 6 spoons |
|---|---|---|---|---|
| Most |  |  |  |  |
| Some |  |  |  |  |
| None |  |  |  |  |

FOSS Next Generation
© The Regents of the University of California
Can be duplicated for classroom or workshop use.

Environments Module
Investigation 3: Brine Shrimp Hatching
No. 15—Notebook Master

## Brine Shrimp–Hatching Conclusions

As a class, we conducted an experiment using multiple trials. Here are the class data on day ____.

**Salt Conditions**

|  | 0 spoons | 2 spoons | 4 spoons | 6 spoons |
|---|---|---|---|---|
| Most |  |  |  |  |
| Some |  |  |  |  |
| None |  |  |  |  |

Write a report answering Dr. Bryan's question. Is salinity an environmental factor that affects brine shrimp egg hatching?

FOSS Next Generation
© The Regents of the University of California
Can be duplicated for classroom or workshop use.

Environments Module
Investigation 3: Brine Shrimp Hatching
No. 16—Notebook Master

## Brine Shrimp–Hatching Conclusions

**WARNING** — This set contains chemicals that may be harmful if misused. Read cautions on individual containers carefully. Not to be used by children except under adult supervision.

As a class, we conducted an experiment using multiple trials. Here are the class data on day ____.

**Salt Conditions**

|  | 0 spoons | 2 spoons | 4 spoons | 6 spoons |
|---|---|---|---|---|
| Most |  |  |  |  |
| Some |  |  |  |  |
| None |  |  |  |  |

Write a report answering Dr. Bryan's question. Is salinity an environmental factor that affects brine shrimp egg hatching?

FOSS Next Generation
© The Regents of the University of California
Can be duplicated for classroom or workshop use.

Environments Module
Investigation 3: Brine Shrimp Hatching
No. 16—Notebook Master

# Response Sheet A—Investigation 3

A student set up a science fair project to find out how temperature affects mealworms. She knew that insects go through different stages in their life cycles. She wanted to know if temperature affected the speed in which the mealworm would go from larva to adult. Below is her lab report.

## Prediction

I think temperature will make the mealworm life cycle go faster. I think that is true because it seems like there are a lot more insects around in the summer when it is warm, than in the winter when it is cold.

## Materials

30 Mealworms

1 Box of wheat bran

3 Cardboard boxes

1 Thermometer

1 Apple

## Procedure

1. Label three boxes: 10°C, 25°C, and 40°C.

2. Put equal amounts of wheat bran in each box (food for the mealworms).

3. Put equal amounts of apple in each box (for moisture).

4. Place 10 mealworms in each box and close the lid.

5. Place each box in an area where you can keep the temperature at the level shown on the box.

6. Check the boxes each day to make sure the mealworms have plenty of wheat bran and apple.

7. Record observations after 2 weeks, 4 weeks, and 6 weeks.

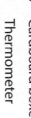

## Data

**After 2 Weeks**

| | 10°C | 25°C | 40°C |
|---|---|---|---|
| Adults | | | |
| Pupae | | 2 | 7 |
| Larvae | 10 | 8 | 3 |

**After 4 Weeks**

| | 10°C | 25°C | 40°C |
|---|---|---|---|
| Adults | | | 5 |
| Pupae | | 4 | 5 |
| Larvae | 10 | 6 | 0 |

**After 6 Weeks**

| | 10°C | 25°C | 40°C |
|---|---|---|---|
| Adults | | | 10 |
| Pupae | | 8 | 0 |
| Larvae | 10 | 2 | 0 |

FOSS Next Generation
© The Regents of the University of California
Can be duplicated for classroom or workshop use.

Environments Module
Investigation 3: Brine Shrimp Hatching
No. 17—Notebook Master

# Response Sheet A—Investigation 3

A student set up a science fair project to find out how temperature affects mealworms. She knew that insects go through different stages in their life cycles. She wanted to know if temperature affected the speed in which the mealworm would go from larva to adult. Below is her lab report.

## Prediction

I think temperature will make the mealworm life cycle go faster. I think that is true because it seems like there are a lot more insects around in the summer when it is warm, than in the winter when it is cold.

## Materials

30 Mealworms

1 Box of wheat bran

3 Cardboard boxes

1 Thermometer

1 Apple

## Procedure

1. Label three boxes: 10°C, 25°C, and 40°C.

2. Put equal amounts of wheat bran in each box (food for the mealworms).

3. Put equal amounts of apple in each box (for moisture).

4. Place 10 mealworms in each box and close the lid.

5. Place each box in an area where you can keep the temperature at the level shown on the box.

6. Check the boxes each day to make sure the mealworms have plenty of wheat bran and apple.

7. Record observations after 2 weeks, 4 weeks, and 6 weeks.

## Data

**After 2 Weeks**

| | 10°C | 25°C | 40°C |
|---|---|---|---|
| Adults | | | |
| Pupae | | 2 | 7 |
| Larvae | 10 | 8 | 3 |

**After 4 Weeks**

| | 10°C | 25°C | 40°C |
|---|---|---|---|
| Adults | | | 5 |
| Pupae | | 4 | 5 |
| Larvae | 10 | 6 | 0 |

**After 6 Weeks**

| | 10°C | 25°C | 40°C |
|---|---|---|---|
| Adults | | | 10 |
| Pupae | | 8 | 0 |
| Larvae | 10 | 2 | 0 |

FOSS Next Generation
© The Regents of the University of California
Can be duplicated for classroom or workshop use.

Environments Module
Investigation 3: Brine Shrimp Hatching
No. 17—Notebook Master

# Response Sheet B—Investigation 3

1. What is the optimum temperature for the mealworms?

   (Circle the one best answer.)

   A. 10°C

   B. 25°C

   C. 40°C

2. What inference can the student make from her observations?

   (Circle the one best answer.)

   A. The warmer the temperature, the faster mealworms move through their life cycle.

   B. All of the mealworms in the 40°C box changed to adults in 6 weeks.

   C. Mealworms in the 10°C box will probably never change into the adult stage.

Record your answers to questions 3 and 4 on the next blank page in your notebook.

3. How do the data show that the student's prediction is right or wrong?

4. Do you think the student should try testing mealworms in temperatures higher than 40°C? Why or why not?

FOSS Next Generation
© The Regents of the University of California
Can be duplicated for classroom or workshop use.

Environments Module
Investigation 3: Brine Shrimp Hatching
No. 18—Notebook Master

# Response Sheet B—Investigation 3

1. What is the optimum temperature for the mealworms?

   (Circle the one best answer.)

   A. 10°C

   B. 25°C

   C. 40°C

2. What inference can the student make from her observations?

   (Circle the one best answer.)

   A. The warmer the temperature, the faster mealworms move through their life cycle.

   B. All of the mealworms in the 40°C box changed to adults in 6 weeks.

   C. Mealworms in the 10°C box will probably never change into the adult stage.

Record your answers to questions 3 and 4 on the next blank page in your notebook.

3. How do the data show that the student's prediction is right or wrong?

4. Do you think the student should try testing mealworms in temperatures higher than 40°C? Why or why not?

Environments Module
Investigation 3: Brine Shrimp Hatching
No. 18—Notebook Master

FOSS Next Generation
© The Regents of the University of California
Can be duplicated for classroom or workshop use.

# Plant Experiment—Water Tolerance

Environmental factor tested:

Controlled environmental factors:

## Procedure

a.  Apply the four container labels and be sure they are securely attached:

b.  Put in 1 level plastic cup of soil without compacting it.

DRY—0 mL
MOIST—0 mL
WET—40 mL
VERY WET—80 mL

c.  Add three seeds of each of the four kinds (12 total). Put the seeds in exactly the same locations in all four containers.

d.  Make sure each seed is 1 cm deep in the soil. Add more soil if needed. Gently pat the soil, using the bottom of the cup.

e.  Add the experimental amount of water evenly to the WET and VERY WET containers, using the 100 mL beaker.

f.  Put each container in a large zip bag and seal it.

g.  Put the bagged container in the FOSS tray.

FOSS Next Generation
© The Regents of the University of California
Can be duplicated for classroom or workshop use.

Environments Module
Investigation 4: Range of Tolerance
No. 19—Notebook Master

# Plant Experiment—Water Tolerance

Environmental factor tested:

Controlled environmental factors:

## Procedure

a.  Apply the four container labels and be sure they are securely attached:

b.  Put in 1 level plastic cup of soil without compacting it.

DRY—0 mL
MOIST—0 mL
WET—40 mL
VERY WET—80 mL

c.  Add three seeds of each of the four kinds (12 total). Put the seeds in exactly the same locations in all four containers.

d.  Make sure each seed is 1 cm deep in the soil. Add more soil if needed. Gently pat the soil, using the bottom of the cup.

e.  Add the experimental amount of water evenly to the WET and VERY WET containers, using the 100 mL beaker.

f.  Put each container in a large zip bag and seal it.

g.  Put the bagged container in the FOSS tray.

FOSS Next Generation
© The Regents of the University of California
Can be duplicated for classroom or workshop use.

Environments Module
Investigation 4: Range of Tolerance
No. 19—Notebook Master

# Plant Experiment—Salt Tolerance

Environmental factor tested:

Controlled environmental factors:

## Procedure

a. Label four planters, one for each type of water.

b. Put 1 level cup of soil in each planter.

c. Put three seeds of the four kinds (12 seeds total) in each planter.

d. Make sure each seed is 1 cm deep in the soil. Add more soil if needed. Gently pat the soil, using the bottom of the cup.

e. Wait for instructions on how to water the seeds.

FOSS Next Generation
© The Regents of the University of California
Can be duplicated for classroom or workshop use.

Environments Module
Investigation 4: Range of Tolerance
No. 20—Notebook Master

**WARNING** — This set contains chemicals that may be harmful if misused. Read cautions on individual containers carefully. Not to be used by children except under adult supervision.

# Plant Experiment—Salt Tolerance

Environmental factor tested:

Controlled environmental factors:

## Procedure

a. Label four planters, one for each type of water.

b. Put 1 level cup of soil in each planter.

c. Put three seeds of the four kinds (12 seeds total) in each planter.

d. Make sure each seed is 1 cm deep in the soil. Add more soil if needed. Gently pat the soil, using the bottom of the cup.

e. Wait for instructions on how to water the seeds.

FOSS Next Generation
© The Regents of the University of California
Can be duplicated for classroom or workshop use.

Environments Module
Investigation 4: Range of Tolerance
No. 20—Notebook Master

## Plant Experiment Setup

Map where each seed is.

**Label**

| Key | |
|---|---|
| | Barley |
| | Corn |
| | Pea |
| | Radish |

Plant all four containers exactly the same.
Environmental factor tested:

Controlled environmental factors:

Number of each seed planted on this date _____ :

Barley _____
Corn _____
Pea _____
Radish _____

FOSS Next Generation
© The Regents of the University of California
Can be duplicated for classroom or workshop use.

Environments Module
Investigation 4: Range of Tolerance
No. 21—Notebook Master

**WARNING** — This set contains chemicals that may be harmful if misused. Read cautions on individual containers carefully. Not to be used by children except under adult supervision.

## Plant Experiment Setup

Map where each seed is.

**Label**

| Key | |
|---|---|
| | Barley |
| | Corn |
| | Pea |
| | Radish |

Plant all four containers exactly the same.
Environmental factor tested:

Controlled environmental factors:

Number of each seed planted on this date _____ :

Barley _____
Corn _____
Pea _____
Radish _____

FOSS Next Generation
© The Regents of the University of California
Can be duplicated for classroom or workshop use.

Environments Module
Investigation 4: Range of Tolerance
No. 21—Notebook Master

# Plant Observations A

Environmental factor tested _____

Planting date _____

Seed type _____

Number of seeds of this kind planted _____

## Part 1. Number of days after planting _____

| Environment | How many plants came up | Height of tallest plant |
|---|---|---|
|  |  |  |
|  |  |  |
|  |  |  |

## Part 2. Number of days after planting _____

| Environment | How many plants came up | Height of tallest plant | Most leaves on one plant |
|---|---|---|---|
|  |  |  |  |
|  |  |  |  |
|  |  |  |  |

FOSS Next Generation
© The Regents of the University of California
Can be duplicated for classroom or workshop use.

Environments Module
Investigation 4: Range of Tolerance
No. 22—Notebook Master

# Plant Observations A

Environmental factor tested _____

Planting date _____

Seed type _____

Number of seeds of this kind planted _____

## Part 1. Number of days after planting _____

| Environment | How many plants came up | Height of tallest plant |
|---|---|---|
|  |  |  |
|  |  |  |
|  |  |  |

## Part 2. Number of days after planting _____

| Environment | How many plants came up | Height of tallest plant | Most leaves on one plant |
|---|---|---|---|
|  |  |  |  |
|  |  |  |  |
|  |  |  |  |

FOSS Next Generation
© The Regents of the University of California
Can be duplicated for classroom or workshop use.

Environments Module
Investigation 4: Range of Tolerance
No. 22—Notebook Master

# Plant Observations B

## Part 3. Number of days after planting _____

| Environment | How many plants came up | Height of tallest plant | | | | |
|---|---|---|---|---|---|---|
| | | | | | | |
| | | | | | | |
| | | | | | | |
| | | | | | | |

| | Most leaves on one plant | Length of longest leaf | Length of longest root |
|---|---|---|---|
| | | | |
| | | | |

**Part 4.** Use this table to determine the range of tolerance. Label the columns. Mark an X where each plant grew.

| | Environment | | | |
|---|---|---|---|---|
| Peas | | | | |
| Corn | | | | |
| Barley | | | | |
| Radishes | | | | |

FOSS Next Generation
© The Regents of the University of California
Can be duplicated for classroom or workshop use.

Environments Module
Investigation 4: Range of Tolerance
No. 23—Notebook Master

---

# Plant Observations B

## Part 3. Number of days after planting _____

| Environment | How many plants came up | Height of tallest plant | | | | |
|---|---|---|---|---|---|---|
| | | | | | | |
| | | | | | | |
| | | | | | | |
| | | | | | | |

| | Most leaves on one plant | Length of longest leaf | Length of longest root |
|---|---|---|---|
| | | | |
| | | | |

**Part 4.** Use this table to determine the range of tolerance. Label the columns. Mark an X where each plant grew.

| | Environment | | | |
|---|---|---|---|---|
| Peas | | | | |
| Corn | | | | |
| Barley | | | | |
| Radishes | | | | |

FOSS Next Generation
© The Regents of the University of California
Can be duplicated for classroom or workshop use.

Environments Module
Investigation 4: Range of Tolerance
No. 23—Notebook Master

## Plant Profile

Plant type _____ Environmental factor _____

Days of growth _____

Label the columns with the environments being studied.

Draw the plants in place.

| | | | | |
|---|---|---|---|---|
| | | Shoot above | | |
| | | Root below | | |

FOSS Next Generation
© The Regents of the University of California
Can be duplicated for classroom or workshop use.

Environments Module
Investigation 4: Range of Tolerance
No. 24—Notebook Master

## Plant Profile

Plant type _____ Environmental factor _____

Days of growth _____

Label the columns with the environments being studied.

Draw the plants in place.

| | | | | |
|---|---|---|---|---|
| | | Shoot above | | |
| | | Root below | | |

FOSS Next Generation
© The Regents of the University of California
Can be duplicated for classroom or workshop use.

Environments Module
Investigation 4: Range of Tolerance
No. 24—Notebook Master

# Response Sheet—Investigation 4

A student designed an experiment to test for the water tolerance of sunflower seeds. She controlled all environmental factors except for the amount of water used. Her design included three water conditions.

Moist: 0 mL of water

Wet: 50 mL of water

Super wet: 100 mL of water

Below are her recorded observations on two different days and her conclusions.

**May 10**

There are 3 sprouts in the moist cup.

**May 13**

There are 3 sprouts in the moist cup and 5 sprouts in the wet cup. One in the wet cup is pretty tall. There is a tall one in the moist cup also. One of the plants in the wet cup has 4 leaves.

## Conclusions

The wet condition is best for growing sunflowers.

Describe how this student could improve her observations.

FOSS Next Generation
© The Regents of the University of California
Can be duplicated for classroom or workshop use.

Environments Module
Investigation 4: Range of Tolerance
No. 25—Notebook Master

---

# Response Sheet—Investigation 4

A student designed an experiment to test for the water tolerance of sunflower seeds. She controlled all environmental factors except for the amount of water used. Her design included three water conditions.

Moist: 0 mL of water

Wet: 50 mL of water

Super wet: 100 mL of water

Below are her recorded observations on two different days and her conclusions.

**May 10**

There are 3 sprouts in the moist cup.

**May 13**

There are 3 sprouts in the moist cup and 5 sprouts in the wet cup. One in the wet cup is pretty tall. There is a tall one in the moist cup also. One of the plants in the wet cup has 4 leaves.

## Conclusions

The wet condition is best for growing sunflowers.

Describe how this student could improve her observations.

FOSS Next Generation
© The Regents of the University of California
Can be duplicated for classroom or workshop use.

Environments Module
Investigation 4: Range of Tolerance
No. 25—Notebook Master

# Teacher Masters

# LETTER TO FAMILY · · · · · · · · · · · · · · · · · · · · · · · · · · · · ·

*Cut here and paste onto school letterhead before making copies.*

------------------------------------------------

======================= **Science News** =======================

Dear Family,

Our class is beginning a new science unit using the **FOSS Environments Module**. We will investigate several different plants and animals to discover the structures and functions that serve in growth and survival and the environments that support their well-being, and attempt to determine the optimum environments for some of the organisms.

The news is full of discussions of the environment—environmental issues, environmental protection, environmental activism, and more. Environmental issues are complex because environments are complex. Our studies will not range into issues, but will deal with the more fundamental question: What is an ecosystem?

Ecosystems can be analyzed in terms of the nonliving components (temperature, moisture, light, and so forth) and living components (organisms). This analysis can take time and often requires close observation and interpretation of results. The interactions among the organisms in food chains and food webs, and the interactions of the organisms with the nonliving environment is the study of ecology. We expect to work with lots of plants and animals to start understanding how organisms flourish when provided with an environment that is just right for them.

Watch for Home/School Connection sheets that I will be sending home from time to time. The activities described on them suggest ways you and your child can extend the environmental inquiry into your home, neighborhood, and community. If possible, take a family field trip to a local aquatic environment to see what lives there and to ponder the environmental factors that affect those organisms there. At another time, you might set up a specialized environment in your home to raise brine shrimp, or embark on a minisafari, looking into microenvironments to see what insects and their kin are living nearby. Perhaps your discoveries will start some family discussions about environments and ecosystems.

We're looking forward to weeks of fun with organisms and their environments! You can get more information on this module by going to www.fossweb.com. If you have questions or comments, or have expertise you would like to share with the class, please drop me a note.

Sincerely,

FOSS Next Generation
© The Regents of the University of California
Can be duplicated for classroom or workshop use.

Environments Module
Investigation 1: Environmental Factors
No. 1—Teacher Master

# MEALWORM LARVA

Environments Module
Investigation 1: Environmental Factors
No. 2 — Teacher Master

# MEALWORM ADULT (DARKLING BEETLE)

Environments Module
Investigation 1: Environmental Factors
No. 4 — Teacher Master

# CRITTER REPLICATOR INSTRUCTIONS . . . . . . . . . . . . . . . .

## Materials for one Replicator

1 Manila file folder, 20 cm × 30 cm

1 *Critter Replicator Cover* sheet

1 Set of body-part-wheel transparencies (head, thorax, abdomen, wings)

4 Round-head paper fasteners, 1/2"

## Construction

1. Glue the cover sheet to the file folder, lining up the bottom of the sheet with the file folder as shown. Cut the folder to the size of the title sheet.

2. Open the folder, and lay it flat with the cover sheet facing up. Cut out the window in the cover sheet. Fold the folder in half, and make a tiny slit at each end of the four paper-fastener locations through all the layers of paper. The slit allows you to easily push the paper fastener through the sheet.

3. Cut out the transparent body-part wheels *inside* the black outline so that the edge of the wheel is clear. In the center of the wheel, cut a hole the width of the paper fastener (not too big!), so that the wheel will turn freely on the fasteners.

4. Cut out the crescent-shaped pieces at the top, bottom, and two sides of the cover sheet and file folder.

5. Use the cover sheet to guide you in positioning each wheel; put the thorax wheel in first and the wing wheel in last. Place each wheel in the folder and lock it in place with a paper fastener through the center. The paper fasteners should go through both sides of the folder.

6. Each body-part wheel should turn freely on its fastener. If a wheel binds, loosen the paper fastener a bit.

FOSS Next Generation
© The Regents of the University of California
Can be duplicated for classroom or workshop use.

Environments Module
Investigation 1: Environmental Factors
No. 6—Teacher Master

# CRITTER REPLICATOR

## Head

### How to use the replicator:

1. Turn all four wheels so the window shows no body parts.
2. Locate an animal from the leaf litter.
3. Observe the animal closely.
4. On your wheel dial the head that most closely resembles the head of the animal.
5. Next select a thorax, the middle part of the animal. How many legs does the animal have?
6. Then choose an abdomen, the last part of the animal.
7. Does the animal have wings?

## Thorax

## Wings

### How to make a record of your critter:

1. Draw your critter on a record sheet.
2. Color your picture to make it more closely resemble the animal you caught.

## Abdomen

FOSS Next Generation
© The Regents of the University of California
Can be duplicated for classroom or workshop use.

Environments Module
Investigation 1: Environmental Factors
No. 7 — Teacher Master

# CRITTER REPLICATOR TRANSPARENCY —HEAD AND THORAX

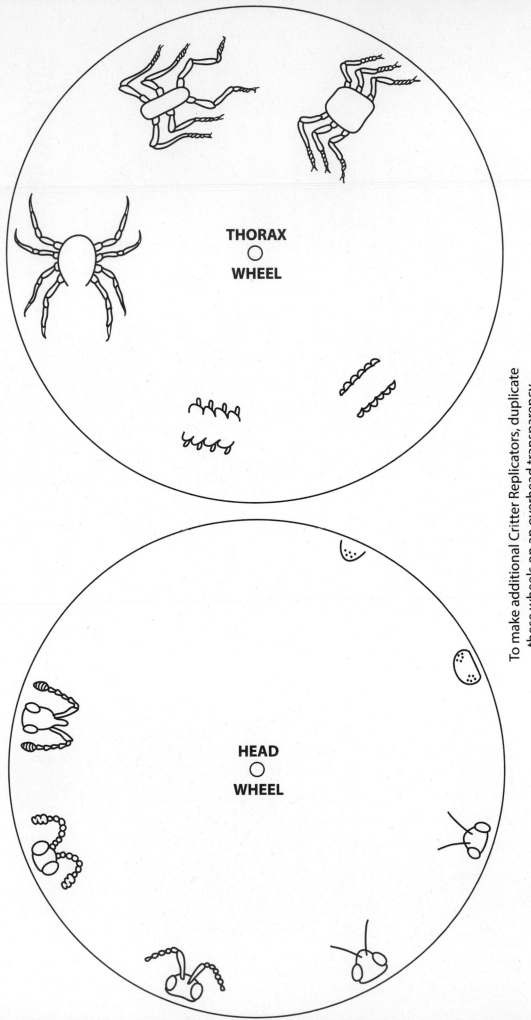

**THORAX**

○

**WHEEL**

**HEAD**

○

**WHEEL**

To make additional Critter Replicators, duplicate these wheels on an overhead transparency.

Environments Module
Investigation 1: Environmental Factors
No. 8 — Teacher Master

FOSS Next Generation
© The Regents of the University of California
Can be duplicated for classroom or workshop use.

WING

○

WHEEL

ABDOMEN

○

WHEEL

To make additional Critter Replicators, duplicate these wheels on an overhead transparency.

Name _____  Date _____

# MATH EXTENSION—PROBLEM OF THE WEEK
## Investigation 1: Environmental Factors

1. A girl made a ladybug cage. Her cage has five sides. Each side has a length of 28 centimeters (cm). The cage is 20 cm high. What is the shape of her cage's base called? What is the perimeter of the base?


2. A boy wants to build rectangular beetle cages to study beetle behavior. He wants all the sides on the base of his cages to be at least 10 cm long. He also wants the base area to equal 576 cm$^2$.

   The boy needs help to find all the possible lengths and widths for the base of the cages. He used only whole centimeters to measure, with no fractional parts.

   What are all the possible dimensions for the rectangular base?

   The boy has decided to make the height of each cage 10 cm. If he fills the container to capacity, how much soil will he need for each cage?

FOSS Next Generation
© The Regents of the University of California
Can be duplicated for classroom or workshop use.

Environments Module
Investigation 1: Environmental Factors
No. 10—Teacher Master

# HOME/SCHOOL CONNECTION

## Investigation 1: Environmental Factors

Insects live just about everywhere. Go on a safari with an adult in and around your home or into the neighborhood to look for insects. You might discover insects that are fun and interesting to observe, like butterflies, bees, moths, grasshoppers, and ladybugs. And, you might see insects that are less welcome, but equally interesting, like ants, flies, mosquitoes, and roaches.

You might also encounter some small animals that are not insects, but live in the same environments, like centipedes, millipedes, isopods, spiders, and worms.

**Safety note.** While most insects, spiders, and other small animals are harmless, some can sting (ants, wasps, bees), and some can bite (spiders, centipedes). Observe the animals without touching them (for your safety) and without disturbing them (for their safety and well-being).

Organize the results of your safari in one of two ways.

- If you find only one or two critters, identify them and describe their environments as best you can. Try to identify both living and nonliving factors in the organisms' environments.

- If you go to an environment and find a number of insects and other small animals living there, write a description of the environment (for example, dark, moist, cool). Then list or draw all the different animals living there.

FOSS Next Generation
© The Regents of the University of California
Can be duplicated for classroom or workshop use.

Environments Module
Investigation 1: Environmental Factors
No.11—Teacher Master

# CLASS AQUARIUM LOG

Start Date _____

| Date | Group responsible | Water temp. | Water level | Were the fish fed? | Observations |
|------|-------------------|-------------|-------------|--------------------|--------------|
|      |                   |             |             |                    |              |
|      |                   |             |             |                    |              |
|      |                   |             |             |                    |              |
|      |                   |             |             |                    |              |

Name _____  Date _____

# MATH EXTENSION—PROBLEM OF THE WEEK

## Investigation 2: Ecosystems

A student wants to set up a tropical-fish aquarium. She has $20.00 to spend on the fish. The store has four types of fish for her tank. She wants at least one of each type of fish. There is no tax charged on fish. She wants to have $2.00 or less left after she buys all of her fish.

| Type of fish | Cost per fish | Length of fish |
|---|---|---|
| Angelfish | $2.98 | 7 cm |
| Lampeye | $1.59 | 3 cm |
| Mollies | $1.35 | 4 cm |
| Neon tetras | $1.70 | 2 cm |

a. What combination of fish could she buy? How much money will she have left? Show all your work.

b. The student's parents agreed to buy the aquarium tank for her new fish. The student remembers from her aquatic-environments project that tropical fish need 1 liter of water for every 3 centimeters (cm) of fish length in the aquarium. What size aquarium in full liters do her fish need? Show your work.

## Bonus problem

Can you find another combination of fish the student could buy? What size tank does she need for these fish?

FOSS Next Generation
© The Regents of the University of California
Can be duplicated for classroom or workshop use.

Environments Module
Investigation 2: Ecosystems
No. 13—Teacher Master

# HOME/SCHOOL CONNECTION

## Investigation 2: Ecosystems

Aquatic environments are homes to fascinating and diverse organisms. Aquatic systems can be as grand as a lake or ocean, or as ordinary as a mud puddle or neglected bucket of water. If you have the opportunity, visit an aquatic environment, like a beach, pond, stream, or canal. Use a collecting net and temporary observation container to see what plants and animals are living in, on, and near the water. Draw pictures to share with the rest of your class. Release the organisms back into the environment and make sure no harm comes to them.

If visiting an aquatic environment is not possible, search around your home and neighborhood for miniaquatic systems (puddles, gutters with standing water, or buckets left out to fill with water). Get down close and look carefully. Maybe scoop some water into a white-bottomed container, such as a half of a milk carton or a plastic bowl.

Describe how one plant or animal is influenced by a **nonliving** factor in the aquatic environment.

_____

_____

_____

Describe how one plant or animal is influenced by a **living** factor in the aquatic environment.

_____

_____

_____

Describe an example of environmental preferences that you observed in the aquatic environment.

_____

_____

_____

If no natural aquatic environments are easy to get to, get a map of the area in which you live. Locate the aquatic environments nearby. List them in groups based on the kind of aquatic environments you locate: stream environments, lake environments, and so forth. Use the reverse side of this paper to make your lists.

FOSS Next Generation
© The Regents of the University of California
Can be duplicated for classroom or workshop use.

Environments Module
Investigation 2: Ecosystems
No. 14—Teacher Master

# CONTAINER LABELS

**Labels for Investigation 3: Brine Shrimp Hatching**

| | |
|---|---|
| **0 spoons of salt** | **2 spoons of salt** |
| **4 spoons of salt** | **6 spoons of salt** |

**Labels for Investigation 4: Range of Tolerance**

| | |
|---|---|
| **MOIST** 0 mL of water | **WET** 40 mL of water |

| | | |
|---|---|---|
| **DRY** 0 mL of water | **SWAMP** 120 mL of water | **VERY WET** 80 mL of water |

**Labels for Investigation 4: Range of Tolerance**

| | |
|---|---|
| **0 spoons of salt** | **1 spoon of salt** |
| **2 spoons of salt** | **4 spoons of salt** |

FOSS Next Generation
© The Regents of the University of California
Can be duplicated for classroom or workshop use.

Environments Module
Investigation 3: Brine Shrimp Hatching
Investigation 4: Range of Tolerance
No. 15—Teacher Master

# ANIMAL SHAPES FOR GROUPS 1–4 ................

## Investigation 3:  Brine Shrimp Hatching

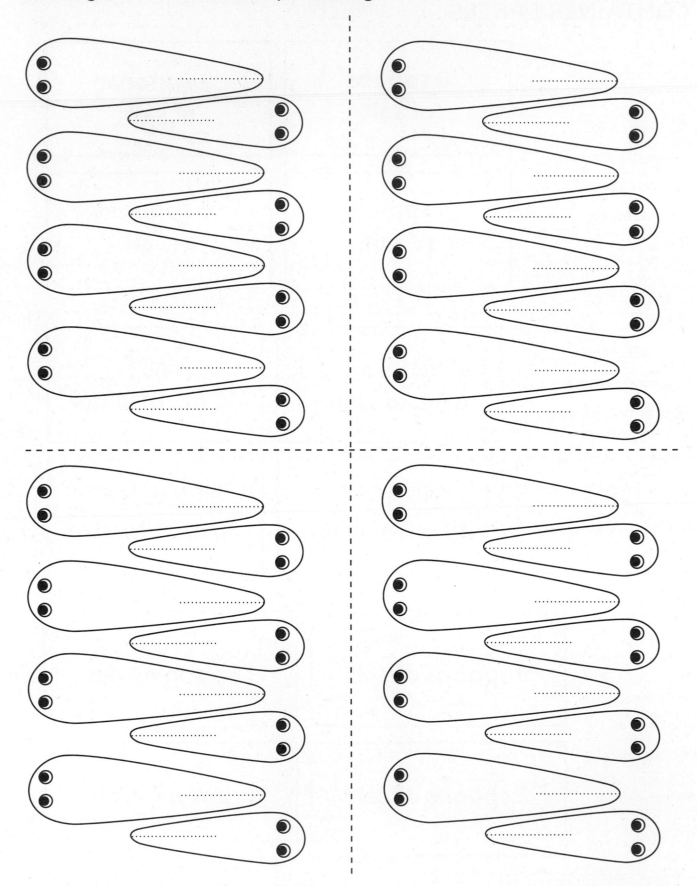

FOSS Next Generation
© The Regents of the University of California
Can be duplicated for classroom or workshop use.

Environments Module
Investigation 3: Brine Shrimp Hatching
No. 16—Teacher Master

Name _____ Date _____

# MATH EXTENSION—PROBLEM OF THE WEEK · · · · · · · ·

## Investigation 3: Brine Shrimp Hatching

A boy wants to set up an experiment to find out the best salt concentration to hatch brine shrimp. He has six containers that hold 0.75 liter of water. He will use a spoonful of brine shrimp eggs per container. He starts with 1/8 of a spoon of salt for the first container.

### Amount of Salt Used

Container 1    1/8 spoon of salt

Container 2    Twice as much salt as he put in container 1

Container 3    Twice as much salt as he put in container 2

Container 4    Twice as much salt as he put in container 3

Container 5    Twice as much salt as he put in container 4

Container 6    Twice as much salt as he put in container 5

How much salt does the boy need for all six containers? Show all your work. Use drawings or tables to help you.

A girl set up a series of six experiments. She wanted to find out the best salt concentration for hatching brine shrimp. She put 1 liter of water and 1 little spoon of brine shrimp eggs in six containers. Then she added a different amount of salt to each container.

### Amount of Salt Used

Container 1    8 spoons of salt

Container 2    Half as much salt as she put in container 1

Container 3    Half as much salt as she put in container 2

Container 4    Half as much salt as she put in container 3

Container 5    Half as much salt as she put in container 4

Container 6    Half as much salt as she put in container 5

How much salt did the girl need for all six containers? Show all your work. Use drawings or tables to help you.

FOSS Next Generation
© The Regents of the University of California
Can be duplicated for classroom or workshop use.

Environments Module
Investigation 3: Brine Shrimp Hatching
No. 17—Teacher Master

# HOME/SCHOOL CONNECTION

## Investigation 3: Brine Shrimp Hatching

Brine shrimp live in salt lakes (like the Great Salt Lake or Mono Lake) or ocean bays (like San Francisco Bay). In their natural environment, the mix of salts is fairly complex. The mix is certainly more complex than just table salt in water. Biologists who study brine shrimp have come up with a recipe for a more complete saltwater environment for the shrimp.

**Metric units**, using liter beaker and metric measuring spoons

| | |
|---|---|
| 1 liter | Pure water |
| 20 milliliters | Rock salt (or other noniodized salt) |
| 5 milliliters | Epsom salts |
| 2 milliliters | Baking soda |

**English units**, using measuring cup and teaspoon (1 teaspoon is about 5 ml)

| | |
|---|---|
| 4 cups | Pure water |
| 4 teaspoons | Rock salt (or other noniodized salt) |
| 1 teaspoon | Epsom salts |
| 1/2 teaspoon | Baking soda |

Stir up this brew. Let it sit in a container (like a cutoff 2-liter soda bottle) for a day or so to mellow. After the brine shrimp hatch, dump them into the new environment, hatching water and all.

Mark the water level with a piece of tape or a permanent marker. As the sea level goes down (evaporation), bring it back to starting level by adding plain water. Why plain water? Only the water evaporates, not the salts. If you renewed the level with salt water, soon the salt concentration would be too great for the brine shrimp to live.

You can keep the colony in a sunny window. The shrimp will need to eat. So dissolve a tiny pinch (seriously, a very little bit) of yeast in a little water. Put a drop of the solution in the brine shrimp container. Give the whole business a gentle stir to distribute the yeast throughout the environment. You can store the remaining yeast solution in the refrigerator. Feed the brine shrimp a bit of yeast solution every few days. Don't get the water too cloudy.

FOSS Next Generation
© The Regents of the University of California
Can be duplicated for classroom or workshop use.

Environments Module
Investigation 3: Brine Shrimp Hatching
No. 18—Teacher Master

# PLANT PROFILE
## Investigation 4: Range of Tolerance

Plant type _____ Environmental factor _____ Days of growth _____

Label the columns with the environments being studied. Tape the plants in place.

| | | | | |
|---|---|---|---|---|
| | | | | |
| | | **Shoot above** | | |
| | | **Root below** | | |
| | | | | |

FOSS Next Generation
© The Regents of the University of California
Can be duplicated for classroom or workshop use.

Environments Module
Investigation 4: Range of Tolerance
No. 19—Teacher Master

# MATH EXTENSION—PROBLEM OF THE WEEK · · · · · · ·

## Investigation 4: Range of Tolerance

A student needs water for his cabbage garden. On a hillside above the garden there is a spring with water flowing underground. The student built a spring box to collect the underground water for his cabbage. He put a pipe from the spring box to his garden.

**Water trickling through sand and gravel**

**Outlet pipe to cabbage garden**

**Wall with small holes through which spring water can seep**

The spring box needs to fill to the top with water before it will flow into the garden.

On the first day, the water level came up 5 centimeters (cm) in the spring box during the night, and then went down 3 cm during the day. The second day, the water level went up another 5 cm at night and down 3 cm during the day. This continues every night and day.

The spring box measures 20 cm from the bottom to the top of the box where the pipe is attached.

On what day or night will water first flow into the student's cabbage garden?

Show all your work. You can use drawings, a chart, or a number line to help you.

FOSS Next Generation
© The Regents of the University of California
Can be duplicated for classroom or workshop use.

Environments Module
Investigation 4: Range of Tolerance
No. 20—Teacher Master

# HOME/SCHOOL CONNECTION . . . . . . . . . . . . . . . .

## Investigation 4: Range of Tolerance

Work on these problems on another piece of paper.

Four friends want to plant a garden. One of their fathers said they could use an area in the backyard that was a rectangle 8 meters by 4 meters.

## Fair plots

The friends want to divide the area into four equal parts. Each friend has a different way to divide the garden plot fairly. Show at least four different ways the garden could be divided. Which one would you pick for the garden? Why?

## How much area to plant?

The friends have decided on a garden plan. They want to know how much area each person has for planting. One friend calculated and said that each would get 12 square meters to plant. Another said they would each get 8 square meters. A third said they would get 8 meters each. Only one of the friends had the correct answer.

What is the area of each person's piece of the garden?

## A row of marigolds

The friends decided to plant a single row of marigolds along one of the 8-meter sides. They will start the row at one corner of the garden plot and plant to the opposite corner. Marigolds need to be planted 10 centimeters apart. How many marigolds will they need for the row?

## A border for the garden

The friends really like the row of marigolds along one side of the garden. They want to plant marigolds around the whole garden. The border of marigolds will be planted along the outer edges, or the perimeter, of the garden. How many marigolds will they need in all to make a border for the entire garden?

FOSS Next Generation
© The Regents of the University of California
Can be duplicated for classroom or workshop use.

Environments Module
Investigation 4: Range of Tolerance
No. 21—Teacher Master

# Assessment Masters

# Embedded Assessment Notes

| Investigation ____, Part ____ | Date_____ |
| --- | --- |

**Got it!**

**Concept**

**Concept**

**Reflections/Next Steps**

| Investigation ____, Part ____ | Date_____ |
| --- | --- |

**Got it!**

**Concept**

**Concept**

**Reflections/Next Steps**

FOSS Next Generation
© The Regents of the University of California
Can be duplicated for classroom or workshop use.

Environments Module
Embedded Assessment Notes
No. 1—Assessment Master

# Performance Assessment Checklist—Environments

Start date _____

End date _____

| Student names | Investigation 1, Part 2 | | | | | |
|---|---|---|---|---|---|---|
| | Asking questions | Planning and carrying out investigations | Analyzing and interpreting data | Constructing explanations | Engaging in argument from evidence | Cause and effect |
| | | | | | | |
| | | | | | | |
| | | | | | | |
| | | | | | | |
| | | | | | | |
| | | | | | | |
| | | | | | | |
| | | | | | | |
| | | | | | | |
| | | | | | | |
| | | | | | | |
| | | | | | | |
| | | | | | | |
| | | | | | | |
| | | | | | | |
| | | | | | | |
| | | | | | | |
| | | | | | | |
| | | | | | | |

FOSS Next Generation
© The Regents of the University of California
Can be duplicated for classroom or workshop use.

Environments Module
Performance Assessment Checklist
No. 2—Assessment Master

# Performance Assessment Checklist—Environments

Start date _____

End date _____

| Student names | Investigation 3, Part 1 | | | |
| | Planning and carrying out investigations | Developing and using models | Systems and system models | Using mathematics and computational thinking | Scale, proportion, and quantity |
|---|---|---|---|---|---|
| | | | | | |
| | | | | | |
| | | | | | |
| | | | | | |
| | | | | | |
| | | | | | |
| | | | | | |
| | | | | | |
| | | | | | |
| | | | | | |
| | | | | | |
| | | | | | |
| | | | | | |
| | | | | | |
| | | | | | |
| | | | | | |
| | | | | | |
| | | | | | |

# Performance Assessment Checklist—Environments

Start date _____

End date _____

| Student names | Planning and carrying out investigations | Obtaining, evaluating, and communicating information | Analyzing and interpreting data | Constructing explanations | Cause and effect |
|---|---|---|---|---|---|
| | | | | | |
| | | | | | |
| | | | | | |
| | | | | | |
| | | | | | |
| | | | | | |
| | | | | | |
| | | | | | |
| | | | | | |
| | | | | | |
| | | | | | |
| | | | | | |
| | | | | | |
| | | | | | |
| | | | | | |
| | | | | | |
| | | | | | |
| | | | | | |

Investigation 4, Part 1

FOSS Next Generation
© The Regents of the University of California
Can be duplicated for classroom or workshop use.

Environments Module
Performance Assessment Checklist
No. 4—Assessment Master

# Assessment Record—Survey/Posttest

Date _____

| Student names | 1 | 2 | 3 | 4a | 4b | 5 | 6a | 6b | 7 | 8 | 9 | 10 | 11 | 12 |
|---|---|---|---|---|---|---|---|---|---|---|---|---|---|---|
| | | | | | | | | | | | | | | |
| | | | | | | | | | | | | | | |
| | | | | | | | | | | | | | | |
| | | | | | | | | | | | | | | |
| | | | | | | | | | | | | | | |
| | | | | | | | | | | | | | | |
| | | | | | | | | | | | | | | |
| | | | | | | | | | | | | | | |
| | | | | | | | | | | | | | | |
| | | | | | | | | | | | | | | |
| | | | | | | | | | | | | | | |
| | | | | | | | | | | | | | | |
| | | | | | | | | | | | | | | |
| | | | | | | | | | | | | | | |

# Assessment Record—Investigation 1 I-Check

Date _____

| Student names | 1 | 2 | 3 | 4 | 5 | 6 | 7 | 8 | 9 | 10a | 10b |
|---|---|---|---|---|---|---|---|---|---|---|---|
| | | | | | | | | | | | |
| | | | | | | | | | | | |
| | | | | | | | | | | | |
| | | | | | | | | | | | |
| | | | | | | | | | | | |
| | | | | | | | | | | | |
| | | | | | | | | | | | |
| | | | | | | | | | | | |
| | | | | | | | | | | | |
| | | | | | | | | | | | |
| | | | | | | | | | | | |
| | | | | | | | | | | | |
| | | | | | | | | | | | |
| | | | | | | | | | | | |
| | | | | | | | | | | | |
| | | | | | | | | | | | |

# Assessment Record—Investigation 2 I-Check

Date _____

| Student names | 1 | 2 | 3 | 4 | 5 | 6 | 7a | 7b | 8a | 8b | 9 | 10 |
|---|---|---|---|---|---|---|---|---|---|---|---|---|
|  |  |  |  |  |  |  |  |  |  |  |  |  |
|  |  |  |  |  |  |  |  |  |  |  |  |  |
|  |  |  |  |  |  |  |  |  |  |  |  |  |
|  |  |  |  |  |  |  |  |  |  |  |  |  |
|  |  |  |  |  |  |  |  |  |  |  |  |  |
|  |  |  |  |  |  |  |  |  |  |  |  |  |
|  |  |  |  |  |  |  |  |  |  |  |  |  |
|  |  |  |  |  |  |  |  |  |  |  |  |  |
|  |  |  |  |  |  |  |  |  |  |  |  |  |
|  |  |  |  |  |  |  |  |  |  |  |  |  |
|  |  |  |  |  |  |  |  |  |  |  |  |  |
|  |  |  |  |  |  |  |  |  |  |  |  |  |
|  |  |  |  |  |  |  |  |  |  |  |  |  |
|  |  |  |  |  |  |  |  |  |  |  |  |  |
|  |  |  |  |  |  |  |  |  |  |  |  |  |
|  |  |  |  |  |  |  |  |  |  |  |  |  |
|  |  |  |  |  |  |  |  |  |  |  |  |  |
|  |  |  |  |  |  |  |  |  |  |  |  |  |
|  |  |  |  |  |  |  |  |  |  |  |  |  |

# Assessment Record—Investigation 3 I-Check

Date _____

| Student names | 1a | 1b | 2a | 2b | 3a | 3b | 4 | 5 | 6 | 7 | 8 | 9 | 10a | 10b |
|---|---|---|---|---|---|---|---|---|---|---|---|---|---|---|
| | | | | | | | | | | | | | | |
| | | | | | | | | | | | | | | |
| | | | | | | | | | | | | | | |
| | | | | | | | | | | | | | | |
| | | | | | | | | | | | | | | |
| | | | | | | | | | | | | | | |
| | | | | | | | | | | | | | | |
| | | | | | | | | | | | | | | |
| | | | | | | | | | | | | | | |
| | | | | | | | | | | | | | | |
| | | | | | | | | | | | | | | |
| | | | | | | | | | | | | | | |
| | | | | | | | | | | | | | | |
| | | | | | | | | | | | | | | |
| | | | | | | | | | | | | | | |
| | | | | | | | | | | | | | | |
| | | | | | | | | | | | | | | |
| | | | | | | | | | | | | | | |
| | | | | | | | | | | | | | | |
| | | | | | | | | | | | | | | |

FOSS Next Generation
© The Regents of the University of California
Can be duplicated for classroom or workshop use.

Environments Module
Assessment Record
No. 8—Assessment Master

1. Look at the picture of the freshwater aquarium. Which of the following is a *nonliving* factor in this environment?

   *(Mark the one best answer.)*

   ○ **A** Air

   ○ **B** Fish

   ○ **C** Plants

   ○ **D** Snails

2. Which example below describes animals competing for food?

   *(Mark the one best answer.)*

   ○ **A**  A rat ignores a sunflower seed but eats a piece of cheese.

   ○ **B**  A snake eats a mouse, and a lizard eats an insect.

   ○ **C**  A gull steals a fish caught by a brown pelican.

   ○ **D**  A blue jay eats an acorn and then is eaten by an owl.

3. Plants get energy from _____.

   *(Mark the one best answer.)*

   ○ **A**  water

   ○ **B**  fertilizer

   ○ **C**  sunlight

   ○ **D**  nutrients

FOSS Next Generation
© The Regents of the University of California
Can be duplicated for classroom or workshop use.

Environments Module
Survey/Posttest
Page 1 of 7

4. A boy wanted to find the optimum conditions for his chickens to lay eggs. He divided his 20 chickens into 2 groups, X and Y, and recorded their daily care and egg production for 2 weeks.

|  | Number of chickens | Daily care (conditions tested) | | Results after 2 weeks |
|---|---|---|---|---|
|  |  | Chicken feed for group | Calcium for group | Number of eggs laid by group |
| **Group X** | 10 | 1,000 grams | 5 grams | 82 |
| **Group Y** | 10 | 1,300 grams | 10 grams | 104 |

a. The boy did not do a good job designing this investigation. He should have _____.

*(Mark the one best answer.)*

○ **A** used more chickens in each group

○ **B** weighed the chickens each day rather than counting eggs

○ **C** recorded how much water he gave the chickens

○ **D** controlled the amount of feed or the amount of calcium

b. Which of the following would most improve the boy's experiment?

*(Mark the one best answer.)*

○ **F** Observe the chickens over a longer period of time

○ **G** Test each condition one at a time

○ **H** Record data for each chicken individually

○ **J** Repeat the same experiment with different groups of chickens

FOSS Next Generation
© The Regents of the University of California
Can be duplicated for classroom or workshop use.

Environments Module
Survey/Posttest
Page 2 of 7

# SURVEY/POSTTEST
## ENVIRONMENTS

5.  Moisture is an example of _____.

    *(Mark the one best answer.)*

    ○ **A**   an environment

    ○ **B**   a preferred environment

    ○ **C**   an optimum condition

    ○ **D**   an environmental factor

6.  A student planted one radish seed in each of five pots. She used the same amount and same kind of soil in each pot. She put them by the same window. The first graph shows the amount of water she put into each pot each day. The second graph shows the length of each root after 7 days.

    a.  What is the *optimum* amount of water a radish seed needs each day to grow long roots? (Use the graphs to answer.)

    *(Mark the one best answer.)*

    ○ **A**   25 mL

    ○ **B**   20 mL

    ○ **C**   15 mL

    ○ **D**   5 mL

    b.  What is the radish seeds' *range of tolerance* for water? (Use the graphs to answer.)

    *(Mark the one best answer.)*

    ○ **F**   5 mL to 25 mL

    ○ **G**   5 mL to 20 mL

    ○ **H**   10 mL to 25 mL

    ○ **J**   10 mL to 20 mL

FOSS Next Generation
© The Regents of the University of California
Can be duplicated for classroom or workshop use.

Environments Module
Survey/Posttest
Page 3 of 7

7. Study the food web shown here.

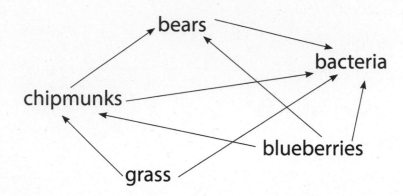

Write **P** in front of the organisms listed below that are *producers*. Write **C** in front of the organisms that are *consumers*. Write **D** in front of the organisms that are *decomposers*.

_____ Bacteria

_____ Bears

_____ Blueberries

_____ Chipmunks

_____ Grass

8. The lateral line on a fish helps the fish sense the presence of other things around it. Which of the following would be a structure with a similar function?

*(Mark the one best answer.)*

◯ **A**  A snail's tentacles

◯ **B**  A dog's fur

◯ **C**  A whale's fins

◯ **D**  An octopus's mouth

lateral line

FOSS Next Generation
© The Regents of the University of California
Can be duplicated for classroom or workshop use.

Environments Module
Survey/Posttest
Page 4 of 7

9. Organisms change over very long periods of time. Giraffes did not always have long necks. Which sentence best explains why giraffes today have long necks?

   *(Mark the one best answer.)*

   ○ **A**  Short-necked giraffes were unable to reproduce.

   ○ **B**  One group of giraffes grew their necks longer to reach higher leaves.

   ○ **C**  Long-necked giraffes were predators of short-necked giraffes.

   ○ **D**  Long-necked giraffes were better suited to the environment.

10. Penguins live in the ocean environment and eat fish. Which of the structures below describes why penguins, a type of bird, can survive in an ocean environment?

    *(Mark the one best answer.)*

    ○ **A**  Penguins have ears that are covered with feathers.

    ○ **B**  Penguins have small wings that act like flippers.

    ○ **C**  Penguins have feathers that absorb large amounts of water.

    ○ **D**  Penguins have a sense of smell that is very limited.

FOSS Next Generation
© The Regents of the University of California
Can be duplicated for classroom or workshop use.

Environments Module
Survey/Posttest
Page 5 of 7

# SURVEY/POSTTEST
## ENVIRONMENTS

• • • • • • • • • • • • • • • • • • • • • • • • • • • • • • • • • • • • • • •

### OPEN-RESPONSE QUESTIONS

11. How do we know that ground sloths and saber-toothed cats (animals that no longer exist on Earth) once lived where Los Angeles, California, is now located? What evidence do we have that allows us to come to this conclusion?

_____

_____

_____

_____

_____

_____

_____

_____

_____

_____

_____

_____

_____

_____

_____

_____

FOSS Next Generation
© The Regents of the University of California
Can be duplicated for classroom or workshop use.

Environments Module
Survey/Posttest
Page 6 of 7

# SURVEY/POSTTEST
## ENVIRONMENTS

## OPEN-RESPONSE QUESTION

12. A class wanted to set up an environment to study hissing cockroaches. A website they found said to feed the cockroaches dry dog food or dry mouse food.

    Write a step-by-step procedure that other students could use to conduct an experiment to determine which kind of food hissing cockroaches prefer.

    _____

    _____

    _____

    _____

    _____

    _____

    _____

    _____

    _____

    _____

    _____

    _____

    _____

    _____

    _____

    _____

    _____

FOSS Next Generation
© The Regents of the University of California
Can be duplicated for classroom or workshop use.

Environments Module
Survey/Posttest
Page 7 of 7

# INVESTIGATION 1 I-CHECK
## ENVIRONMENTS

1.  Write **L** next to each *living* environmental factor. Write **N** next to each *nonliving* environmental factor.

    _____ Fire

    _____ Earthworms

    _____ Wind

    _____ Rocks

    _____ Grass

2.  A student predicts that squirrels prefer sunflower seeds to corncobs. To test her prediction, she puts sunflower seeds in one feeder and corncobs in another feeder. The student hangs the feeders in the same tree and observes the feeders for half an hour each afternoon.

    What is the most important observation this student needs to record to find out which food squirrels prefer?

    *(Mark the one best answer.)*

    ○ **A**   How often the feeders need to be filled

    ○ **B**   The number of squirrels that eat at each feeder

    ○ **C**   How often the same squirrel eats at each feeder

    ○ **D**   How many times each squirrel chases another away from the feeder

3.  Temperature is an example of _____.

    *(Mark the one best answer.)*

    ○ **A**   an environmental factor

    ○ **B**   a preferred environment

    ○ **C**   a controlled variable

    ○ **D**   an environment

FOSS Next Generation
© The Regents of the University of California
Can be duplicated for classroom or workshop use.

Environments Module
Investigation 1 I-Check
Page 1 of 5

Name _____

4. Sanderlings are small shorebirds. They are found on sandy beaches, mudflats, lagoons, and rocky shores. A student decided to study the sanderlings and determine their preferred environment. He spent 15 minutes each day observing each environment and recorded his observations in his science notebook.

| Sanderling Observations (15 minutes/day) | | | | |
|---|---|---|---|---|
| | Sandy beach | Mudflat | Lagoon | Rocky shore |
| Monday | 24 | 12 | 7 | 4 |
| Tuesday | 26 | 8 | 5 | 0 |
| Wednesday | 18 | 15 | 5 | 3 |
| Thursday | 21 | 6 | 4 | 7 |
| Friday | 25 | 9 | 11 | 0 |

Using the data in the table above, which is the Sanderling's preferred environment?

*(Mark the one best answer.)*

○ **A** Sandy beach

○ **B** Mudflat

○ **C** Lagoon

○ **D** Rocky shore

Sanderling

5. Everything that surrounds and influences an organism is its _____.

*(Mark the one best answer.)*

○ **A** environmental factor

○ **B** preference

○ **C** environment

○ **D** basic need

FOSS Next Generation
© The Regents of the University of California
Can be duplicated for classroom or workshop use.

Environments Module
Investigation 1 I-Check
Page 2 of 5

6.  Fish breathe with gills rather than lungs. Isopods breathe with gill-like structures. For this reason, isopods require _____.

    *(Mark the one best answer.)*

    ○ **A**   a sandy environment

    ○ **B**   a moist environment

    ○ **C**   a warm environment

    ○ **D**   a dry environment

7.  Which structure labeled on the mealworm allows the mealworm to move from place to place?

    *(Mark the one best answer.)*

    ○ **A**

    ○ **B**

    ○ **C**

    ○ **D**

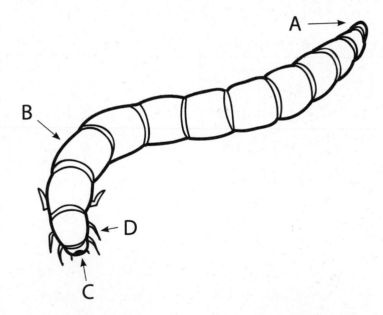

FOSS Next Generation
© The Regents of the University of California
Can be duplicated for classroom or workshop use.

Environments Module
Investigation 1 I-Check
Page 3 of 5

8. A student asks the question "Do isopods prefer dark environments or light environments?" In order to test his question, which setup below should the student use?

*(Mark the one best answer.)*

○ **A**    Start isopods here.

○ **B**    Start isopods here.

○ **C**    Start isopods here.

○ **D**    Start isopods here.

9. What happens if an organism moves outside of its preferred environment?

*(Mark the one best answer.)*

○ **A**    It dies immediately.

○ **B**    It dies within a year, but not immediately.

○ **C**    It survives, but it is not happy.

○ **D**    It survives as long as its basic needs are met.

FOSS Next Generation
© The Regents of the University of California
Can be duplicated for classroom or workshop use.

Environments Module
Investigation 1 I-Check
Page 4 of 5

# INVESTIGATION 1 I-CHECK
## ENVIRONMENTS

· · · · · · · · · · · · · · · · · · · · · · · · · · · · · · · · · · ·

## OPEN-RESPONSE QUESTIONS

10. Humans, like other organisms, get living and nonliving factors from their environment in order to survive.

   a. Name one *living* factor and describe how it allows you to survive in your environment.

   _____

   _____

   _____

   _____

   _____

   _____

   _____

   b. Name one *nonliving* factor and describe how it allows you to survive in your environment.

   _____

   _____

   _____

   _____

   _____

   _____

   _____

   _____

FOSS Next Generation
© The Regents of the University of California
Can be duplicated for classroom or workshop use.

Environments Module
Investigation 1 I-Check
Page 5 of 5

1. A community of organisms interacting with one another and with nonliving factors in an area is called _____.

   *(Mark the one best answer.)*

   ○ **A**   an environment

   ○ **B**   a population

   ○ **C**   an ecosystem

   ○ **D**   a habitat

2. Mark an **X** next to each organism that makes its own food from resources in the environment.

   _____   A blade of grass

   _____   An earthworm

   _____   An insect

   _____   An oak tree

   _____   A cheetah

3. Write **M** next to the words that describe *matter* in an ecosystem. Write **E** next to the words that describe *energy* in an ecosystem. Write **ME** next to words that describe both matter and energy in an ecosystem.

   _____   air

   _____   sunlight

   _____   food

   _____   water

FOSS Next Generation
© The Regents of the University of California
Can be duplicated for classroom or workshop use.

Environments Module
Investigation 2 I-Check
Page 1 of 6

4. Different sense receptors are specialized for particular kinds of information gathering. Animals have many amazing adaptations that allow them to survive in their environment.

   Match each of the information-gathering animal adaptations with its function.

   *(Wrtie a letter in each blank.)*

   _____ The star-nosed mole has 100,000 nerve fibers that run from its nose to its brain. (That's more nerve fibers than you have on your hands.)

   _____ A hawk's vision is 20/5. (Normal vision for people is 20/20. A hawk can see at 20 feet what you see at 5 feet from the object.)

   _____ Dolphins and bats can bounce sound waves off of objects. They can hear up to at least 100,000 Hz. (Humans only hear in the 1,000 to 20,000 Hz range.)

   _____ Crabs and crayfish have hairs on their claws and other body parts that can sense water current and vibration.

   _____ Dogs have a membrane for smelling that is 150 sq cm (compared to humans whose smell membrane is 4 sq cm!).

   _____ Pigeons have eyes mounted on either side of their heads. This allows them a field of vision of 340°—that's all but a small part of a full circle!) (Humans on the other hand, with their eyes mounted forward have only a 190° field.)

**This adaptation is good for**

A   Feeling around for food in the dark

B   Seeing a predator coming from any direction

C   Navigating in water or air where vision is not helpful

D   Sensing a change in the environment when danger is near

E   Understanding the environment by its smell

F   Being able to sight prey from far away

FOSS Next Generation
© The Regents of the University of California
Can be duplicated for classroom or workshop use.

Environments Module
Investigation 2 I-Check
Page 2 of 6

5. Mark an **X** next to each sentence that explains what most likely happens if an environment changes *slightly*.

_____ All living organisms die.

_____ Some animals move to new locations.

_____ Plants and animals can grow and reproduce faster.

_____ Some plants and animals survive and reproduce.

_____ Organisms change their structures to survive in the new conditions.

6. The table shows a deer population over several years. From these data, what would you predict is the carrying capacity for deer in this ecosystem?

*(Mark the one best answer.)*

○ **A**   30 to 35 deer

○ **B**   142 deer

○ **C**   5 to 30 deer

○ **D**   5 to 142 deer

| Year | Number of deer |
|------|----------------|
| 1 | 5 |
| 2 | 35 |
| 3 | 142 |
| 4 | 30 |
| 5 | 32 |

FOSS Next Generation
© The Regents of the University of California
Can be duplicated for classroom or workshop use.

Environments Module
Investigation 2 I-Check
Page 3 of 6

7. Study the food web shown here.

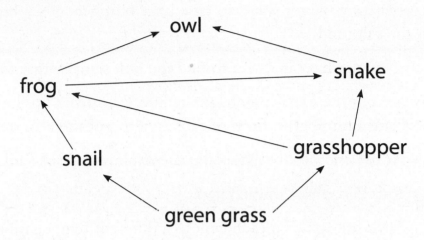

a. What type of organism is missing from this food web?

   *(Mark the one best answer.)*

   ○ **A**  Producers

   ○ **B**  Consumers

   ○ **C**  Decomposers

   ○ **D**  Carnivores

b. If the frog population suddenly increased, which organism would be most likely to decrease right away?

   *(Mark the one best answer.)*

   ○ **F**  Owls

   ○ **G**  Snakes

   ○ **H**  Leaves

   ○ **J**  Snails

FOSS Next Generation
© The Regents of the University of California
Can be duplicated for classroom or workshop use.

Environments Module
Investigation 2 I-Check
Page 4 of 6

# INVESTIGATION 2 I-CHECK
## ENVIRONMENTS

• • • • • • • • • • • • • • • • • • • • • • • • • • • • • • • • • • • • • • • • • • •

8. Daphnia are small animals that can live in water as cold as 5°C and as warm as 30°C. A pet store owner wants to raise large numbers of daphnia to sell to customers for fish food.

The pet store owner has hired you to find the best temperature for raising daphnia.

a. You wrote out the experimental procedure, but mixed up the order of the steps. Write a number in front of each step to put them in the correct order.

_____ At the end of 2 weeks, count the daphnia in each tank.

_____ Put 25 daphnia in each tank.

_____ Heat the water in each tank to a different temperature: 5°C, 10°C, 15°C, 20°C, 25°C, 30°C.

_____ Observe daily, and record observations.

_____ Set up six identical tanks of water. Put 1 liter (L) of water in each container.

b. From which step will you use the data to make your recommendation to the pet store owner?

*(Mark the one best answer.)*

○ **A** At the end of 2 weeks, count the daphnia in each tank.

○ **B** Put 25 daphnia in each tank.

○ **C** Observe daily, and record observations.

○ **D** Set up six identical tanks of water.

FOSS Next Generation
© The Regents of the University of California
Can be duplicated for classroom or workshop use.

Environments Module
Investigation 2 I-Check
Page 5 of 6

# INVESTIGATION 2 I-CHECK
## ENVIRONMENTS

## OPEN-RESPONSE QUESTIONS

9. Explain how energy from the Sun helps a shark survive.

   (Hint: Think of a food chain.)

   _____

   _____

   _____

   _____

   _____

   _____

   _____

10. Describe how an aquatic ecosystem is similar to a terrestrial ecosystem and how it is different.

    *The ecosystems are similar because* _____

    _____

    _____

    _____

    _____

    *The ecosystems are different because* _____

    _____

    _____

    _____

    _____

FOSS Next Generation
© The Regents of the University of California
Can be duplicated for classroom or workshop use.

Environments Module
Investigation 2 I-Check
Page 6 of 6

1. A student studied brine shrimp hatching under a range of saltwater conditions. He organized his results on the graph you see here.

**Brine Shrimp Hatching in Salt Water**

Estimated number of brine shrimp hatched

Number of spoons of salt in 1 liter (L) of water

a. Which condition is the *optimum* for hatching brine shrimp?

*(Mark the one best answer.)*

○ **A**  0 to 10 spoons of salt

○ **B**  5 spoons of salt

○ **C**  3 to 7 spoons of salt

○ **D**  7 spoons of salt

b. What is the *range of tolerance* for hatching brine shrimp?

*(Mark the one best answer.)*

○ **F**  7 spoons of salt

○ **G**  0 to 7 spoons of salt

○ **H**  3 to 7 spoons of salt

○ **J**  5 spoons of salt

FOSS Next Generation
© The Regents of the University of California
Can be duplicated for classroom or workshop use.

Environments Module
Investigation 3 I-Check
Page 1 of 7

2. A kangaroo rat has a very long tail, has small front legs, varies in color from dark gray to light brown, and has fur-lined cheek pouches in its mouth. It uses the pouches to store food, which also serves as the rat's source of water.

    a. Which adaptation of the kangaroo rat enables it to survive in the desert?

    *(Mark the one best answer.)*

    ○ **A**  Very long tails

    ○ **B**  Small front legs

    ○ **C**  Variation in fur color from dark gray to light brown

    ○ **D**  Mouth pouches for storing food

    b. Which adaptation would help a population of kangaroo rats survive if it were moved to an environment with dark soil rather than light-colored soil?

    *(Mark the one best answer.)*

    ○ **F**  Very long tails

    ○ **G**  Small front legs

    ○ **H**  Variation in fur color from dark gray to light brown

    ○ **J**  Mouth pouches for storing food

FOSS Next Generation
© The Regents of the University of California
Can be duplicated for classroom or workshop use.

Environments Module
Investigation 3 I-Check
Page 2 of 7

Name _____

**A Mountain Environment**

| Temperature | This side gets a lot of rain. | This side doesn't get much rain. |
|---|---|---|
| -10°C to 0°C | A | |
| 0°C to 10°C | | |
| 10°C to 20°C | | B |
| 20°C to 30°C | | |
| 30°C to 40°C | | |

▲ Pine tree   ♣ Oak tree

3. Study the diagram of a mountain.

   a. What one factor probably causes the lack of pine trees in Area B?

   *(Mark the one best answer.)*

   ○ **A**  The temperature in Area B is too cold.

   ○ **B**  Area B doesn't get enough sunlight.

   ○ **C**  The air in Area B doesn't have enough oxygen.

   ○ **D**  Area B doesn't get much rain.

   b. What is the temperature range of tolerance for *oak* trees?

   *(Mark the one best answer.)*

   ○ **F**  −10°C to 40°C

   ○ **G**  0°C to 40°C

   ○ **H**  20°C to 40°C

   ○ **J**  30°C to 40°C

FOSS Next Generation
© The Regents of the University of California
Can be duplicated for classroom or workshop use.

Environments Module
Investigation 3 I-Check
Page 3 of 7

4. What is the relationship between range of tolerance and the optimum for an environmental factor?

*(Mark the one best answer.)*

- ○ **A**  The optimum is within the range of tolerance.
- ○ **B**  The range of tolerance is within the optimum.
- ○ **C**  Range of tolerance and optimum are not related.
- ○ **D**  Range of tolerance and optimum are the same thing.

5. A fertilizer company releases leftover products into a river environment. What is most likely to happen to the organisms living in that river environment?

*(Mark the one best answer.)*

- ○ **A**  The plants will benefit and get healthier, but animals will most likely die.
- ○ **B**  All of the organisms will become healthier.
- ○ **C**  Some organisms will die, and others will benefit from the change.
- ○ **D**  All of the plants and animals will die.

FOSS Next Generation
© The Regents of the University of California
Can be duplicated for classroom or workshop use.

Environments Module
Investigation 3 I-Check
Page 4 of 7

6. What organism is missing from this food web?
   *(Mark the one best answer.)*

   ○ **A**   Phytoplankton

   ○ **B**   Octopus

   ○ **C**   Lobster

   ○ **D**   Small fish

7. Organisms of the same kind differ a little bit in how they look and behave because _____.

   *(Mark the one best answer.)*

   ○ **A**   the environment determines how they look and act

   ○ **B**   there is variation within one kind of organism

   ○ **C**   inherited information is exactly the same from parent to offspring

   ○ **D**   organisms adapt to whatever surroundings they are placed in

8. The rabbit population in a certain ecosystem sometimes decreases dramatically. One possible explanation for this decrease is _____.

   *(Mark the one best answer.)*

   ○ **A**   an increase in the field mouse population

   ○ **B**   a decrease in the owl population

   ○ **C**   an increase in the grass population

   ○ **D**   an increase in the coyote population

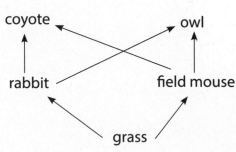

FOSS Next Generation
© The Regents of the University of California
Can be duplicated for classroom or workshop use.

Environments Module
Investigation 3 I-Check
Page 5 of 7

# INVESTIGATION 3 I-CHECK
**ENVIRONMENTS**

## OPEN-RESPONSE QUESTIONS

9.  A student read on the Internet that if you add a chemical called BigFish to the water in your aquarium, the fish will grow twice as fast. Unfortunately, the article did not say how much of the chemical should be added each week.

Design an experiment that this student could use to find out how much chemical to add to his 80-liter (L) tank to grow bigger fish.

_____

_____

_____

_____

_____

_____

_____

_____

_____

_____

_____

_____

_____

_____

_____

FOSS Next Generation
© The Regents of the University of California
Can be duplicated for classroom or workshop use.

Environments Module
Investigation 3 I-Check
Page 6 of 7

# INVESTIGATION 3 I-CHECK
## ENVIRONMENTS

. . . . . . . . . . . . . . . . . . . . . . . . . . . . . . . . . . . . . .

### OPEN-RESPONSE QUESTIONS

10. You have closely observed several different organisms in class: mealworms, isopods, goldfish, guppies, and brine shrimp. All of these organisms look very different, which makes them easy to tell apart. But if you look even more closely, you might have seen that two mealworms don't look exactly alike, even if they are the same type of organism.

   a. Describe how one of the organisms you studied, isopods for example, varies from one to another.

   _____

   _____

   _____

   _____

   _____

   _____

   _____

   b. Describe why this variation might be important to a population's survival.

   _____

   _____

   _____

   _____

   _____

   _____

   _____

FOSS Next Generation
© The Regents of the University of California
Can be duplicated for classroom or workshop use.

Environments Module
Investigation 3 I-Check
Page 7 of 7